Decision Making in Complex Task Environments

Decision Modes in Complex Task Environments

Norbert Steigenberger
Thomas Lübcke
Heather M. Fiala
Alina Riebschläger

CRC Press
Taylor & Francis Group
Boca Raton London New York

CRC Press is an imprint of the
Taylor & Francis Group, an **informa** business

CRC Press
Taylor & Francis Group
6000 Broken Sound Parkway NW, Suite 300
Boca Raton, FL 33487-2742

© 2017 by Taylor & Francis Group, LLC
CRC Press is an imprint of Taylor & Francis Group, an Informa business

No claim to original U.S. Government works

Printed on acid-free paper

International Standard Book Number-13: 978-1-1387-4846-0 (Paperback)

Library of Congress Cataloging-in-Publication Data

Names: Steigenberger, Norbert, author.
Title: Decision modes in complex task environments / Norbert Steigenberger, Heather M. Fiala, Thomas Lübcke, Alina Riebschläger.
Description: Boca Raton, FL : CRC Press, 2017.
Identifiers: LCCN 2016052150 | ISBN 9781138748460 (paperback : alk. paper) | ISBN 9781498796620 (ebook)
Subjects: LCSH: Decision making. | Search and rescue operations--Decision making--Case studies.
Classification: LCC BF448 .S74 2017 | DDC 658.4/03--dc23
LC record available at https://lccn.loc.gov/2016052150

Visit the Taylor & Francis Web site at
http://www.taylorandfrancis.com

and the CRC Press Web site at
http://www.crcpress.com

Contents

Acknowledgments

We gratefully acknowledge financial support received from the German Federal Ministry of Education and Research under grant numbers 13N12106, 13N12107, 13N13011, and 13N13014. In addition, we are grateful for the continuous support we received during both research projects from our advisor, Martin Bettenworth, from *vdi Technologiezentrum*.

We also wish to express our gratitude toward the professionals in the maritime search and rescue services in the Baltic and North Sea region who allowed us in-depth insights into their work and who filled questionnaires and took part in interviews. In particular, we acknowledge the excellent cooperation with the Royal Danish Navy, the Danish Naval Homeguard, and the German Maritime Search and Rescue Service.

Authors

Dr. Norbert Steigenberger is an assistant professor at Jönköping International Business School (Sweden). Previously, he held researcher and guest researcher positions at the University of Cologne (Germany) and Aalto University Helsinki (Finland). He studied Business Economics at the Technical University Freiberg (Germany) and the Universitetet i Tromsø (Norway) and received his doctorate from Chemnitz Technical University in 2011. His research primarily concerns different topics in management and organization, such as decision-making in organizations, mergers and acquisitions, and novel forms of digital business.

Thomas Lübcke is head of research and development at the German Maritime Search and Rescue Service (DGzRS). He studied sociology and education at the University of Rostock (M.A.). His current research addresses organizational issues in complex maritime search and rescue operations and transdisciplinary research on human factors in safety critical domains as well as in the simulation and training of interorganizational working contexts. As an active member, he is also involved in the platform "People in Complex Work Environments."

Dr. Heather M. Fiala is a professional staff member of the Center for Controlling & Management at the Institute of Management Accounting and Control of the WHU—Otto Beisheim School of Management in Vallendar, Germany. She studied psychology at Beloit College in Beloit, Wisconsin (B.A.) and the University of Erfurt (M.Sc.) in Germany. She completed her PhD in psychology in 2013 at the University of Erfurt. From 2012 to 2015, she worked as a researcher at the Seminar for Business Administration, Corporate Development, and Organization at the University of Cologne, investigating decision-making in complex task environments.

Alina Riebschläger studied business administration at the University of Cologne. From 2013 to 2015, she was a graduate assistant at the Seminar for Business Administration, Corporate Development, and Organization at the University of Cologne and assisted in a research project concerning decision-making in complex task environments.

1 Why Decision Mode Matters

1.1 TWO DEATHS ON CHRISTMAS EVE

It's Christmas Eve and a violent storm is raging.* The rescue cruiser *Eugen Hermann* left port in the late afternoon to aid another ship in distress. After several taxing hours, the rescue mission was successful and the crew is looking forward to being back on shore. However, on their way back to port, the crew and cruiser struggle to cope with the bad sea state. While coxswain, Anders Mueller, and engineer, Osvald Kettil, are situated on the open bridge, the other two crew members remain below deck. Due to the stormy weather, the entire crew is wearing survival suits and life vests and is fastened to the guide rails with safety lines. At 10:14 p.m., still a long distance from the safe harbor, the maritime rescue coordination center (MRCC) on shore loses radio contact with the cruiser after receiving a mayday call.

So what happened? At 10:14 p.m., the rescue cruiser is hit by ground swell. A rogue wave, a so-called *kaventsmann*, rolls the cruiser along its longitudinal axis. These huge, unpredictable waves can arise when waves from deeper water run onto shallow coastal areas. Due to its construction, the cruiser rights itself again but has suffered serious damage. Engineer Osvald Kettil, who had briefly unfastened his safety line to check the engines, is gone. Two other crew members are hurt, the ship disabled, the mast broken, and communication devices dead.

Alarmed by the mayday call and the interrupted radio contact, the MRCC sends a rescue helicopter, which reaches the scene at 11:45 p.m., and requests additional aid from other rescue cruisers. Still, it will be several more hours before those vessels reach the last known position of the *Eugen Hermann*. The remaining crew members, who were all saved by their safety lines, see the helicopter approaching and head to the cruiser's forecastle, attempting to reach the lifesaving rescue sling being lowered by the helicopter. Unfortunately, due to the rough sea and resulting list of the cruiser, their attempts fail. Coxswain Anders Mueller thus decides that the crew should return to the relatively safe deckhouse. When reentering the ship, the crew members must momentarily remove their safety lines from the guide rail. Tragically, a second breaker wave hits the vessel at the exact moment when Anders Mueller, the last crew member to enter the deckhouse, is no longer affixed to the guide rail. He, too, is lost. The helicopter is powerless to provide aid in the rough weather conditions. Several hours later, the two rescue cruisers dispatched earlier in the night arrive at the scene and finally manage to successfully rescue the two surviving crew members and tow the damaged cruiser back to harbor. The bodies of Osvald Kettil and Anders Mueller

---

* Based on a true event. Names have been changed.

are recovered several days later. The vessel is later repaired and modernized before returning to service along with one of the surviving crew members. The second survivor remains permanently unfit for duty.

During this incident, two decisions crucially influenced the course of disaster and led, leading more or less directly to the fatalities. First, the engineer decided to remove his safety line from the guardrail in order to check the engines. Second, the coxswain led his crew to the cruiser's forecastle in an attempt to reach the helicopter's rescue sling and then halted the attempt in favor of returning to the deckhouse. Both decisions were, in hindsight, wrong and led to the fatal outcome. Although we will never know with certainty whether the deaths could have been avoided, it is at least reasonable to speculate that different decisions would have led to different and, potentially, less fatal outcomes. These critical decisions were extremely difficult to make; search and rescue professionals act in a demanding and potentially dangerous environment almost every day, yet only rarely do they find themselves in such immediate danger. Hence, the crew members had little previous experience in this exact type of situation on which they could base their decisions. In addition, the decision environment was highly complex: a combination of darkness, stormy weather, and a level of danger that was difficult to assess made it extremely difficult for the decision-makers to accurately predict the outcome of any decision. Both decisions had no ex ante optimal solution but were instead probabilistic. That is, the decision-makers could, at best, base their decisions on assumptions of how likely a positive outcome would be for each decision option. However, they had no way of ascertaining whether a certain decision would lead to a specific outcome with certainty. In all instances, there was no "best" decision.

Probabilistic decisions, although not always as critical as in the *Eugen Hermann* case, are quite typical in organizations engaged in complex task environments. These decisions are burdened with uncertainty and other sources of complexity that go beyond the domain of well-structured selection problems and reach far into the field of judgments. These judgmental decisions are not about finding the objectively correct answer. Instead, they involve the development of a strong judgment regarding which option has a good probability of producing a positive outcome based on the information that is available and comprehensible at the time that the decision is made as well as with the knowledge that even high probabilities of success do not guarantee a positive outcome. In each decision situation described above, there would have been good arguments for all options that were possible at the specific time. Inspecting the engines was important; it was probably worth taking a risk. Walking the crew to the fore and then back again was a risk as well, but it could have meant a way to safety. Still, the decision-makers did not include, or perhaps misinterpreted, important cues that could have possibly prevented them from making the decisions that led to the deaths of two crew members on Christmas Eve.

In many organizational environments, decisions tend to be of the judgmental type, in which an ideal situation cannot be reached with certainty and decision-makers must base their judgments on incomplete information. Oftentimes, such decisions tend to be invisible and go unnoticed for both the decision-maker and casual onlookers. If the rogue wave had not hit the cruiser at the exact time that it did, taking the engineer with it, Osvald Kettil would have survived and likely remained unaware of the fact that he had just made an important decision when he removed his safety line from the guide rail to go below deck; in his mind, he may simply have been doing his job, checking the engines. It is only in some cases that such decisions draw our conscious awareness.

Would the outcome have been better if Kettil would have first taken a moment to carefully consider his options before removing his safety line instead of acting quickly, presumably based on his work routine? Perhaps, but that would have disturbed his work flow, potentially costing him important time, and may have gone against his preferred method of making decisions. Would he have kept his safety line on if he had had more experience with that type of situation? Probably. Still, experience is not always readily available. This is especially true in the case of extreme and unusual situations. The same considerations can be made for the coxswain's decisions. Again, we cannot know what would have happened if he had, for example, continued the evacuation attempts or not attempted rescue via helicopter at all and, instead, kept his crew in the relative safety of the deckhouse.

Typically, decisions made in complex task environments (which we will define more precisely in Section 1.3) do not have such dramatic consequences, and as stated above, many decisions might not be recognizable as decisions at all. Indeed, a large portion of professionals' performance is based on intuitive decision-making—that is, decisions made without the decision-maker's conscious awareness. This allows professionals to make complex decisions quickly and effectively, leading to high decision-making and work performance in both time-critical and routine work conditions. Examples of this type of decision are those made in action in military, firefighting, and police work, where time pressure is sometimes so strong that actors do not have time to consider alternatives or weigh options. Consider, for example, tactical decision-making during a fast-paced police operation, during which decisions must be made in milliseconds, or the decision-making of firefighters in a burning structure. In such cases, decisions often have to be made *immediately*, leaving no time for conscious consideration. Expertise-based intuition allows police officers and firefighters in such situations to effectively reach good decisions that foster the specific tactical goals of an assignment.

For some decisions, on the other hand, organizations actively discourage this type of intuitive decision-making and instead steer employees toward conscious, deliberate analysis. A good example is the use of checklists in aviation cockpits. Many pilots are very familiar with their planes and might have a good sense for issues that may arise. Still, research and practice have shown that these feelings are inferior to systematic checks in that field, meaning that flight safety is best served with a highly systematic, sequential, and structured approach to the critical decision of whether a plane is fit for flight. Accordingly, organizational structures in these situations have been designed such that the decision to initiate the flight requires the pilot to actively consider an extensive list of potential malfunctions by checking a variety of on-board systems for proper functionality.

Many situations in complex task environments fall somewhere in between these two polar categories. As a result, decision-makers have substantial control over the degree to which they use each decision mode—deciding either quickly and intuitively or, instead, contributing cognitive effort and time into a decision by switching to a deliberate mode. Both decision modes differ substantially regarding underlying cognitive processes and, accordingly, have very different advantages and disadvantages, as we will discuss later in this chapter.

If we take a step back at this point and look at the field of decision-making more broadly, we will see that scholars from many disciplines, such as psychology, economics, neuroscience, and political science, have examined decision-making for decades and even centuries—for very good reason. Decisions shape human action as well as the outcome of work and social processes. Everyone has rich stories to tell of situations in which they made poor decisions with potentially negative consequences or when they came to a surprising and, at first glance, illogical decision that helped them reach a good outcome in a difficult situation. Whereas the earlier centuries of decision-making research concerned themselves mostly with tools to reach logical and ideal decisions under certain conditions, in more recent decades the focus has shifted more and more toward the cognitive perspective of decision-making: How are people actually making decisions; and how and why does this (oftentimes) work? The curious ability of the human brain to quickly arrive at complex decisions in highly demanding situations, in connection with the astonishing abilities that humans display regarding analytical information processing, has piqued the interest of scholars (Bernstein, 1996; Dane & Pratt, 2007). Overall, we have come a long way in understanding how decision-making works on a cognitive level. However, we still know very little in some specific fields, which will be the main focus of this book—more specifically, regarding the question of when decision-makers approach decisions with deliberate thought or, instead, trust their gut feelings. Considering the substantial differences in underlying cognitive mechanisms, structural preconditions, subjective and objective cognitive load, decision biases, and likely outcomes, this question is of considerable importance for our understanding of how humans make decisions in complex task environments.

Based on the notion that individual performance and well-being as well as organizations and societies and, not seldom, even life itself hinges on the effectiveness of human decision-making, understanding how and why humans make decisions the way they do is and will remain an important cornerstone of psychology and the social sciences as well as our understanding of human performance in and beyond organizations in general. This book is our modest contribution to that body of knowledge.

1.2 LOGIC AND STRUCTURE OF THIS BOOK

Research on human decision-making has a long tradition, which we outline briefly in Section 2.1. Over the decades, we have arrived at a relatively clear understanding of how decision-making works at a cognitive level. In particular, one model of human cognition has developed that found widespread support in psychological science (Evans & Stanovich, 2013) and also forms the basis of our understanding of the cognitive foundations of human decision-making: Dual-processing theory. According to dual-processing theory, human cognition and decision-making consists of both a quick, automated, and holistic information processing system and a deliberate, slow, and analytical system that interact in specific ways (Evans, 2008). In Section 2.2, we will discuss these two distinct types of information processing and their relationship to each other in more detail. The fast and automated system leads to what is typically called intuitive decision-making whereas the analytical system leads to deliberate decision-making. In real-life decision situations, decision-makers typically apply some combination of these archetypical approaches. As outlined, both intuition and deliberation have different advantages and disadvantages. Thus, it matters which decision mode a decision-maker applies in a given decision situation. For example, the deliberate mode allows for the application of complex rules as well as the inclusion of additional information and is robust toward decision biases. However, this decision mode is also slow and loads heavily on limited cognitive resources (Kahneman & Frederick, 2002; Kurzban, Duckworth, Kable, & Myers, 2013). In real-life complex task environments, in which decision-makers might face different tasks simultaneously and cognitive resources may be stretched thin, these disadvantages could weigh heavily. In comparison, intuitive decision-making has a much larger information processing capacity, is much faster, and allows access to information that is not available to conscious thought. Still, intuition crucially hinges on tacit learning and can be subject to severe decision biases, meaning that intuitive judgments are often wrong or heavily biased if the decision-maker cannot draw on relevant experience (Dijksterhuis, 2004; Hogarth, 2010; Tversky & Kahneman, 1974). In Sections 2.3 and 2.4, we will present an in-depth analysis of the strengths and weaknesses of both deliberate and intuitive decision-making.

Based on these insights, it becomes clear that decisions look quite differently, depending on whether a decision-maker used a largely deliberate or intuitive decision strategy, as both have different strengths and weaknesses. In order to understand organizational and individual-level outcomes of the choice of decision modes, it is therefore helpful to understand why decisions that led to these outcomes were made the way we observed them. In addition, whereas we have a relatively clear understanding of the conditions that favor effective intuition or deliberation, our understanding of *when* decision-makers *actually* apply intuition or deliberation in real-life complex task environments is limited. In Section 2.5, we will outline what is known about the selection of decision modes, leading to the insight that our knowledge is substantially incomplete, especially regarding insights into real-life decision-making.

The decisions people make and their outcomes can be observed under experimental conditions in traditional laboratory studies typically applied in psychological research. From this stream of work, we have obtained the bulk of our understanding regarding how intuition and deliberation function on a cognitive level as well as when intuition and deliberation are likely successful (e.g., Tversky & Kahneman, 1974). However, these studies tell us little about how decision-makers select decision modes in complex, real-life tasks, where the stakes are high and situations are far from controlled. Accordingly, this type of research has limited predictive power for understanding real-life decision-making in complex task environments and requires complementary research in real-life decision situations (Lipshitz, Klein, Orasanu, & Salas, 2001a). In particular, observing the processes that lead decision-makers to favor one decision mode over another in real-life environments is difficult for various reasons. First, as various conditions might affect this choice simultaneously, such as a need for information, uncertainty, and time pressure, establishing causality is difficult in both laboratory settings and real life. If decisions are also important, issues such as stress and mental pressure—which are difficult or even impossible to recreate in a laboratory setting—might also become relevant. Second, decision modes are also difficult to observe as processes that happen in the decision-maker's mind are not easily captured. Combinations of observational and self-reporting measures are required to develop scientifically robust data on which additional analyses can be based.

Still, if we do not understand why decision-makers decide the way they do in important real-life decisions, we are a long way from understanding the performance of decision-makers and organizations operating in complex, high-stakes environments and, as a result, cannot contribute to improving this performance. Obtaining such an understanding would help us develop recommendations for decision-makers that design decision (work) environments with the goal of further improving the performance of organizations in complex task environments.

The empirical part of this book, Chapters 3 through 5, picks up at this point. Based on a variety of studies conducted in a collaborative research effort between the Department of Corporate Development and Organization at the University of Cologne (Germany) and the German Maritime Search and Rescue Service (DGzRS), we will address in detail several conditions that affect the choice of decision mode in complex task environments. Our empirical approach involves survey and video-based studies, action research, and expert interviews. We collected data in real-life work environments, large-scale training exercises, and in-house training simulators. Triangulating these methodological approaches and empirical settings provides us with the opportunity to develop strong empirically grounded insights into decision modes in complex task environments with a focus on maritime search and rescue and relevance far beyond this narrow professional delineation. We will describe both the empirical field and the studies we conducted in Chapter 3. In Chapter 4, we will report the results we obtained from several survey studies conducted over three years in the field of maritime search and rescue. We will present insights on the contingency conditions that we found to affect the choice of decision modes and also discuss a variety of conditions that had no or only an inconsistent effect in our research. These insights will then be integrated into the current state of knowledge as outlined in Chapter 2. We will also briefly discuss how the choice of decision mode affects a decision-maker's satisfaction with a decision. Chapter 5 adds to these insights with a qualitative perspective. We will report on two shadowing studies. The first follows decision-makers through large maritime exercises, whereas the second covers the initial stages of a rescue mission in the Aegean Sea, during which a group of professionals, including one of the authors of our book, had to develop solutions for novel tasks in the context of rescuing refugees sailing from the coast of Turkey to Greece. In these studies, we combined video-based observational methods with expert focus interviews and elements of action research. Chapter 5 will provide qualitative insights into how decision-making is embedded in organizations, how decision patterns develop, and how experience, in particular, affects these processes.

The concluding Chapter 6 will then summarize, discuss, and synergize these multifaceted insights in order to formulate recommendations and practical implications with a particular focus on the conditions that can be utilized to improve decision-making in complex task environments. We will develop general recommendations for decision-maker trainings in complex task environments and suggest a specific training approach to help decision-makers choose an appropriate decision mode in complex situations. Moreover, we will suggest a small research agenda to further improve our understanding of decision-making in complex task environments from an academic perspective.

This book rests empirically on a specific setting: maritime search and rescue. The perils of the sea have haunted seafarers since humankind decided to build ships to brave the waters. Thus, it does not come as a surprise that organized sea rescue has a long tradition. The establishment of the first modern-day sea rescue station in 1776 by William Hutchinson at Formby Point near Liverpool may be considered the beginning of modern maritime search and rescue. In the decades that followed, the idea of an organized sea rescue service gained traction across Europe. For example, the German Maritime Search and Rescue Service, the focal organization in large portions of the empirical studies presented in this book, was established in 1865. The rescue of people in distress at sea and castaways, the search and rescue of missing persons, and the salvage of people killed in maritime accidents are some of the responsibilities of maritime search and rescue services, which are just as important today as they were two or three hundred years ago. Whereas the general mission of lifeboat organizations has not changed substantially over the past centuries, the specific challenges are in a constant state of development. Currently, the growing prevalence of offshore energy production and cruise ship traffic as well as mass migration overseas are some of the important challenges encountered by maritime search and rescue organizations.

Maritime search and rescue is what has been called a high-reliability task environment, comparable to firefighting, aviation, police work, military operations, and the operation of complex technical installations, in which decisions potentially affect the lives and health of others or have the potential to substantially impair the natural environment (Weick, Sutcliffe, & Obstfeld, 1999). Studies targeting real-life decision-making have traditionally focused on high-reliability organizations (Helsloot & Groenendaal, 2011; Lipshitz, Klein, Orasanu, & Salas, 2001b; Zsambok, 1997); and this book follows with that tradition. An important reason for choosing this field to study decision modes in comparison to, for example, decision-making in business, is that decisions tend to be clearly visible and linkable to specific outcomes, which is often not the case in more ambiguous organizational environments. Another reason to focus on high-reliability organizations it their obvious importance for our societies; decisions made in high-reliability organizations tend to directly affect the lives and health of others. Although high-reliability organizations are a specific form of organization, in line with previous reasoning (Gore, Banks, Millward, & Kyriakidou, 2006), we claim that research on decision-making in high-reliability contexts offers substantial insights for other fields in which critical decisions are made, which is the case in many workplaces and life environments. We will discuss the degree of applicability of our insights to other settings at various points throughout this book.

The combination of a thorough review of the literature on decision modes and several empirical studies is our attempt to aid the scientific and professionals community to go one step further in the direction of understanding how professionals make complex real-life decisions in challenging environments. Overall, the aim of this book is to substantially advance knowledge on decision modes in complex task environments based on rich empirical insights obtained in the field of maritime search and rescue.

The organizational background for the fruitful collaboration between an academic and a professionals institution were two research projects conducted between 2012 and 2017. We gratefully acknowledge financial support for both research projects from the German Federal Ministry of Education and Research, which allowed us to conduct the empirical studies we report in this book. In the context of this research, we were able to collect original data that is unique in terms of both breadth and depth.

1.3 DEFINING THE FIELD: DECISIONS IN COMPLEX TASK ENVIRONMENTS

At the outset of this book, it is important to delineate our focus of interest and establish definitions for key concepts that we will consistently use in the following chapters.

1.3.1 DECISIONS IN COMPLEX TASK ENVIRONMENTS

In this book, we focus on a specific type of decision situation, which we call decisions in complex task environments. Specifically, we examine important, real-life, judgmental decisions for which decision-makers must deal with incomplete information and uncertainty and rely on their own best judgments to quickly arrive at a decision. These decisions have no best solution. Rather, at the time of decision-making, it is not certain which outcome each decision option will produce. Decision-makers do not always face very high time pressure, but there is some constraint on the time a decision-maker has, which might be somewhere between several seconds and some hours. Decisions of this kind are quite common in everyday life as well as in many work environments such as, for example, medicine, aviation, management, and many other professional fields. More formally, we are interested in *judgmental decisions* that are characterized by *relevance*, *complexity*, and at least moderate *time pressure* and occur in *real-life settings* (in contrast to artificial laboratory settings). Each of these characteristics has specific implications for decision-making.

1.3.2 JUDGMENTAL DECISIONS

Judgmental decision situations have no ex ante objectively ideal solution. Instead, decision-makers perceive and evaluate a set of cues and base their decisions on how they think these cues relate to possible decision options and resulting outcomes (Dane & Pratt, 2009). This means that, at the time of decision-making, it is not possible for a decision-maker to be certain that a specific decision option is indeed the best choice in a given situation. Instead, they must identify options and derive probability judgments to make a promising choice. This is due to the element of uncertainty that characterizes these decisions and concerns the relationship between cues that could inform a decision and potential outcomes. That is, in a judgmental decision of the type we examine here, a decision-maker cannot be absolutely certain that cue A always leads to outcome B. There may be other conditions in place that affect this cue–outcome relationship. It is also possible that the relationship itself is unstable in that judgmental decision situations tend to be ill-structured and it is often not clear at the outset how a specific cue relates to a decision outcome (cf. Kahneman & Klein, 2009). For example, the perception of smoke coming from a window would be such a decision cue. It could refer to a serious fire or just burnt toast. In judgmental decisions, decision-makers must evaluate the cues as they perceive them and then select a decision option based on their best judgment of a specific situation.

1.3.3   RELEVANCE AND REAL-LIFE ENVIRONMENTS

We focus on situations in which a wrong decision could have substantial negative consequences for the decision-maker, other persons, society, or the natural environment. Examples are decisions in high-stakes work environments such as medicine, search and rescue, and aviation (Weick & Sutcliffe, 2001; Weick et al., 1999). In typical laboratory studies, in which a substantial portion of previous research on decision-making is based, this notion of relevance is absent. In laboratory settings, test persons typically solve artificial decision tasks that have no intrinsic meaning to them and for which success and failure bear little to no consequence. These artificial situations offer the advantage of experimental control but also have obvious limitations when it comes to predicting real-life behavior (Lipshitz et al., 2001a). Thus, we focus instead on decisions in situations in which people act and work as they do in real life—that is, in which task outcomes matter for the decision-maker and the tasks themselves are meaningful and within the decision-maker's field of experience (Gore et al., 2006). In such situations, decision-makers are under varying degrees of pressure to perform well—a pressure that is also partially absent in laboratory settings. The second aspect that differentiates our approach (which is in some respects comparable to the approach of the naturalistic decision-making research stream, which we discuss in more detail in Section 2.1) from traditional laboratory studies is that we examine relevant decisions of professionals, that is, well-educated individuals, in their work environments. These individuals have contextualized knowledge that they can draw upon as well as work routines that potentially affect decision-making to a substantial degree (see Huang & Pearce, 2015). We discuss the critical role of experience more thoroughly in Chapter 2. In contrast, laboratory research relies heavily (yet not exclusively) on observations of decision-making behavior of undergraduate students (Lipshitz et al., 2001a) and also often ignores the importance of contextualized knowledge and work routines.

1.3.4   COMPLEXITY

We define complex decisions as multicue decisions. In these situations, decision-makers select decision options based on multiple, heterogeneous, and potentially incomplete cues and face uncertainty regarding the focal situation and the relationship between cues and potential outcomes. The notion of complexity is closely related to the nature of a judgmental decision, meaning that a judgmental decision always includes some aspects of complexity due to the inherent uncertainty that defines a decision as judgmental. In these decision situations, decision-makers evaluate perceived cues and base their decisions on their best judgment of how cues relate to outcomes. The perceived cues are a "lens" (Newell & Shanks, 2014, p. 2) through which a decision-maker experiences a situation. The cues thus mediate between an ongoing event and decision-makers' perceptions thereof. This means that a decision-maker will not directly perceive an event but, rather, only (some) cues related to that event and must then develop a mental picture of the event based on the perceived cues. A decision-maker on board an approaching search and rescue vessel might perceive smoke emanating from a window of another vessel but be left unaware regarding the source and location of the fire—for example, whether the smoke is related to a substantial fire or bad cooking—key information required to develop an ideal solution. Experience and training help decision-makers interpret cues in a manner that facilitates good decision-making.

What specifically determines the subjective complexity of a decision depends on the focal situation. However, two general dimensions are of particular importance for the complexity of multicue judgments: the perceptibility of cues (cf. Söllner, Bröder, & Hilbig, 2013) and the stability of the relationship between each cue and situational outcomes (Kahneman & Klein, 2009). The ease with which a decision-maker can perceive the relevant cues depends on characteristics of the situation. It is, for example, much more difficult to spot a person in the water or interpret the cause of a fire at night than it is during the daytime. Complexity is high if there are few cues or if cues are difficult to spot but also if there is a large number of cues, which may or may not be relevant for the decision—for example, when a decision-maker is confronted with many people who all provide potentially inconclusive and/or contradictory accounts of an event. Environmental conditions, such as visibility and sea state, thus critically affect complexity (Norrington, Quigley, Russell, & van der Meer, 2008). Decisions in maritime search and rescue made during fog, for example, will be much more difficult to make than in clear weather, as many pieces of information that could inform a decision cannot be easily perceived. The second aspect relates closely to the nature of judgmental decisions. If cues tend to be misleading or only erratically related to outcomes—that is, if the uncertainty that governs the relationship between cue and the related event is high—complexity is also high. Furthermore, situational dynamics might lead a decision situation to develop in unexpected directions, adding a general element of surprise, which is also an element of situational complexity. An extreme example would be the occurrence of the *kaventsmann*, the rogue wave that claimed the life of the engineer in the case outlined at the beginning of this chapter.

1.3.5 TIME PRESSURE

Finally, time pressure is an important issue in most complex task environments. As stated previously, decisions must be made within a certain time frame, which might be seconds, minutes, or hours. If a decision must be made within minutes or an even shorter time frame, it often may not be possible to think through all decision options and assess and compare all possible actions. Thus, deliberate decision-making cannot be applied unlimitedly. The most critical aspect of time pressure is that it implies opportunity costs for the application of deliberation; time and energy spent on active deliberation is not available for other tasks that might also be important such as making other decisions, engaging in action, communicating, or resting. Time pressure might be enforced by situational dynamics, for example, a fire spreading or persons drowning or, more prosaically, by workload, when new demands appear frequently so that the time that a decision-maker can spend on a focal task or decision is limited.

In this book, we will focus on decisions that adhere to these four characteristics. These are the decisions that we will call *decisions in complex task environments*.

REFERENCES

Bernstein, P. L. 1996. *Against the gods: The remarkable story of risk*. New York: John Wiley & Sons.

Dane, E., & Pratt, M. G. 2007. Exploring intuition and its role in managerial decision making. *Academy of Management Review*, 32(1): 33–54.

Dane, E., & Pratt, M. G. 2009. Conceptualizing and measuring intuition: A review of recent trends. *International Review of Industrial and Organizational Psychology*, 24: 1–40.

Dijksterhuis, A. 2004. Think different: The merits of unconscious thought in preference development and decision making. *Journal of Personality and Social Psychology*, 87(5): 586–598.

Evans, J. S. B. T. 2008. Dual-processing accounts of reasoning, judgment, and social cognition. *Annual Review of Psychology*, 59(1): 255–278.

Evans, J. S. B. T., & Stanovich, K. E. 2013. Dual-process theories of higher cognition: Advancing the debate. *Perspectives on Psychological Science*, 8(3): 223–241.

Gore, J., Banks, A., Millward, L., & Kyriakidou, O. 2006. Naturalistic decision making and organizations: Reviewing pragmatic science. *Organization Studies*, 27(7): 925–942.

Helsloot, I., & Groenendaal, J. 2011. Naturalistic decision making in forensic science: Toward a better understanding of decision making by forensic team leaders. *Journal of Forensic Sciences*, 56(4): 890–897.

Hogarth, R. M. 2010. Intuition: A challenge for psychological research on decision making. *Psychological Inquiry*, 21(4): 338–353.

Huang, L., & Pearce, J. L. 2015. Managing the unknowable: The effectiveness of early-stage investor gut feel in entrepreneurial investment decisions. *Administrative Science Quarterly*, 60(4): 634–670.

Kahneman, D., & Frederick, S. 2002. Representativeness revisited: Attribute substitution in intuitive judgment. In T. Gilovich, D. Griffin, & D. Kahneman (Eds.), *Heuristics and biases: The psychology of intuitive judgment*: 49–81. Cambridge University Press.

Kahneman, D., & Klein, G. 2009. Conditions for intuitive expertise: A failure to disagree. *American Psychologist*, 64(6): 515–526.

Kurzban, R., Duckworth, A., Kable, J. W., & Myers, J. 2013. An opportunity cost model of subjective effort and task performance. *The Behavioral and Brain Sciences*, 36(6): 661–679.

Lipshitz, R., Klein, G., Orasanu, J., & Salas, E. 2001a. A welcome dialogue—And the need to continue. *Journal of Behavioral Decision Making*, 14(5): 385–389.

Lipshitz, R., Klein, G. A., Orasanu, J., & Salas, E. 2001b. Taking stock of naturalistic decision making. *Journal of Behavioral Decision Making*, 14(5): 331–352.

Newell, B. R., & Shanks, D. R. 2014. Unconscious influences on decision making: A critical review. *The Behavioral and Brain Sciences*, 37(1): 1–19.

Norrington, L., Quigley, J., Russell, A., & van der Meer, R. 2008. Modelling the reliability of search and rescue operations with Bayesian belief networks. *Reliability Engineering & System Safety*, 93(7): 940–949.

Söllner, A., Bröder, A., & Hilbig, B. E. 2013. Deliberation vs. automaticity in decision-making: Which presentation format features facilitate automatic decision making? *Judgment and Decision Making*, 8(3): 278–298.

Tversky, A., & Kahneman, D. 1974. Judgment under uncertainty: Heuristics and biases. *Science*, 185: 1124–1131.

Weick, K. E., & Sutcliffe, K. M. 2001. *Managing the unexpected: Assuring high performance in an age of complexity* (1st ed.). San Francisco: Jossey-Bass.

Weick, K. E., Sutcliffe, K. M., & Obstfeld, D. 1999. Organizing for high reliability: Processes of collective mindfulness. *Research in Organizational Behaviour*, 21: 81–123.

Zsambok, C. E. 1997. Naturalistic decision-making: Where are we now? In C. E. Zsambok & G. Klein (Eds.), *Naturalistic decision-making*: 3–16. Mahwah, NJ: Erlbaum.

2 Perspectives on Decision-Making in Complex Task Environments

What We Do and Do Not Know

2.1 A BRIEF HISTORY OF THE SCHOLARLY DEBATE ON DECISION-MAKING IN COMPLEX ENVIRONMENTS

In order to develop a sound understanding of the phenomenon of human decision-making in complex tasks, it is helpful to briefly talk about the roots of our current understanding of decision-making in complex situations. At the outset of this chapter, we will therefore highlight some key developments that decision-making research underwent in the last decades and centuries in order to derive a clearer understanding of the challenges we currently face and address in the empirical studies presented in this book. Although such a historical review must necessarily remain incomplete due to the complexity and sheer size of the decision-making research field, it helps us understand the main traditions from which our understanding of decision-making in complex tasks stems.

Research on decision-making is an inherently interdisciplinary endeavor with a long tradition in diverse disciplines: Psychology, economics, and various social sciences have long studied decision-making processes and outcomes, applying their respective conceptual and empirical perspectives, theories, and methodologies. The study of risk and choice (see Bernstein, 1996) might be seen as the origin of modern-day debates on decision-making research. This research field originated in the works of scholars in the 18th century, in particular that of Daniel Bernoulli (1700–1782), who aligned studies of risk and probability with considerations of the decision-maker's preferences. In this classical tradition of decision-making research, which forms the foundation for large portions of economic theories concerning value maximization, the decision-maker is what has been called an "Economic Man": This normatively ideal decision-maker is fully informed, responds effectively to environmental cues, and is fully rational in a sense that neither affective experience nor limited cognitive capacity interfere with decision-making. In addition, he or she makes decisions such that the expected utility of a decision is maximized (Edwards, 1961). Based on this perspective, a strong normative research stream on decision-making as value maximization has emerged that is grounded in the notion that decision-makers strive for objectively optimal decisions that maximize payoffs—a notion that has attracted much critique over the decades for being unrealistic (e.g., Einhorn & Hogarth, 1981). In real life, decision-makers are typically unable and often unwilling to engage in this type of normatively ideal decision-making behavior. This is due to the fact that cognitive capacity and time are limited; decisions often lack an ideal outcome—that is, they are uncertain; and information is not always readily available. Given the limited ability of the classical approach to describe human decision-making in many relevant real-life situations, new perspectives arose in the late 1940s and 1950s that rejected the notion of perfect rationality and perfect information, adding instead time pressure, cost of information acquisition, and social interactions to the conceptual framework. In particular, Herbert Simon laid the foundation for a perspective that remains highly influential for today's understanding of decision-making in stable environments in organizations (Puranam, Stieglitz, Osman, & Pillutla, 2015). Simon suggested that decision-making in organizations can be understood as being "boundedly rational," meaning that a decision-maker will typically strive for "rational" (meaning objectively good) decisions but, at the same time, be restricted in his or her search for optimal decisions by limited access to information, organizational embedment, and limited processing capacity. Accordingly, the goal of the boundedly rational decision-maker is to identify a decision option that satisfies his or her respective aspiration level instead of one that achieves an objectively ideal outcome (Simon, 1955).

Deviating further from the concept of "rational" utility maximization and continuing the journey toward a descriptive theory of decision-making, scholars such as Ward Edwards and Paul Meehl advanced what has come to be known as the behavioral decision theory (BDT) and the judgment and decision-making (JDM) stream in the 1950s and 1960s. Those perspectives forwarded concepts of subjective evaluation of probability (instead of objective evaluations of probability) and other variables related to the decision-maker (Edwards, 1961) to describe how decision-makers arrive at their decisions. The roots of behavioral decision theory include Simon's (1955) concept of bounded rationality as well as Luce and Raiffa's (1957) investigation of decision-making under uncertainty.

The next leap forward in understanding human decision-making were the contributions of Daniel Kahneman and Amos Tversky in what has come to be known as the heuristics and biases paradigm (e.g., Tversky & Kahneman, 1974), prospect theory (e.g., Tversky & Kahneman, 1979), and decision framing (e.g., Tversky & Kahneman, 1981). With Kahneman, Tversky, and Simon, the notion of intuition as a means to approach decisions without or with only limited conscious deliberation entered the arena in force. Kahneman and Tversky noted that many people did not think consciously about decisions at all but instead referred to mental shortcuts prone to a variety of systematic biases, which Kahneman and Tversky termed "intuition." We will discuss this view on intuition more closely in Sections 2.2 and 2.4.2. Prospect theory—one of the first theories to view the individual not as a rational decision-maker but rather as one prone to cognitive biases—describes individuals' perceptions of possible gains and losses relative to a decision outcome. In contrast to the simplistic idea of utility maximization, which was previously widespread in decision-making research (Bernstein, 1996), prospect theory proposes that individuals subconsciously expect potential losses to hurt more than they expect potential gains to feel good. The implication is that individuals are inherently loss averse and will be more motivated to avoid loss than they are to seek out potential gains (Tversky & Kahneman, 1979). A result of this innate loss aversion—Kahneman and Tversky's third major contribution to decision-making research—is the concept of framing, which states that the way in which a choice is presented (i.e., as a potential gain or loss) affects the decision outcome, as decision-makers prefer, for example, certain gains to probabilistic gains as well as probabilistic losses to certain losses (Tversky & Kahneman, 1981). That is, if individuals are presented with a choice between receiving US$10 for certain or a 10% chance of receiving US$100, most would prefer the certain gain of US$10 even though both options have the same expected value. In turn, confronted with the choice of having to pay US$10 for certain or a 10% chance of paying US$100, most individuals prefer to take the 10% chance. If a choice can be framed as a gain or a loss, then this framing will heavily affect the choice people make even if the underlying decision problem is "objectively absolutely identical." All three contributions (intuition biases, prospect theory, framing) proved to be highly influential for the development of a descriptive perspective on decision-making and, with that, our current understanding of human decision-making in complex task environments.

Although decision research is primarily based on laboratory studies in which decision-makers act in artificial, fully controlled environments, researchers found that the organizational context in which a decision-maker is embedded plays a major role in the decision processes (Puranam et al., 2015; Simon, 1955). Building on that notion, the descriptive perspective on decision-making forwarded by Tversky, Kahneman, and others as well as the observation that experienced decision-makers, in particular, handle highly complex decision tasks rather effectively despite (or rather because of) the fact that they rely on their intuition, a group of scholars took a radically inductive perspective on decision-making in complex, real-life task environments, which came to be known as "naturalistic" decision-making (Orasanu & Connolly, 1995, p. 19). Naturalistic decision-making focuses on highly proficient decision-makers acting in complex decision settings that are characterized by uncertainty, high stakes, ill-structured problems, and time pressure and asks how those individuals are able to decide effectively under those challenging circumstances (Gore, Banks, Millward, & Kyriakidou, 2006). This perspective was a radical departure from the previous intuition-skeptical perspective forwarded by researchers in the tradition of Kahneman and Tversky. Field research is of major importance in the context of naturalistic decision-making, whereas most research in the traditions of psychology and behavioral economics rests on laboratory studies. The core insight of naturalistic decision-making is that experienced decision-makers develop field-specific expertise that they use to subconsciously model the outcome of conceivable decision options, triggered by observations they make (Lipshitz, Klein, Orasanu, & Salas, 2001b). Based on these recognition-primed decision models (Klein, 1995), decision-makers then select the decision option that promises to produce the best solution to a complex problem. As this process of perceiving, mental modeling, and option selection occurs subconsciously, this type of intuitive decision-making is very fast (Kahneman & Klein, 2009). Naturalistic decision-making thus developed insights into effective intuition (as compared to intuition as an error-prone mental shortcut). This approach forwarded a heroic model of an experienced decision-maker who could solve complex tasks effectively and quickly without the need to refer to conscious thought—a perspective that is insightful but also drew critique for both conceptual and methodological reasons. In particular, the extensive use of retrospective interviews as a methodological basis and lack of a formalized theory in naturalistic decision-making has been widely criticized (Lipshitz, Klein, Orasanu, & Salas, 2001a).

At the same time, knowledge of cognitive processes that govern human decision-making has made substantial progress. One foundation of these approaches is the notion that both deliberation and intuition are of importance for human decision-making—often at the same time (Chen & Bargh, 1999; Evans & Stanovich, 2013). Psychologists began to disentangle conceptions of intuition and align it with other important concepts, distinguishing, for example, between mature and immature intuition and aligning the concept of intuition with experience and individual decision-style preferences (Baylor, 2001; C. Betsch, 2008; T. Betsch & Glöckner, 2010). Others studied the abilities and limits of deliberate decision-making, outlining, for example, the role of experience for the effective processing of information in deliberate thought (Dijkstra, Pligt, & Kleef, 2013).

Laboratory-based research has developed a strong understanding of the conditions that favor intuition over deliberation and vice versa—in particular, those contingent upon task context. Today, the advantages and limitations of both deliberate and intuitive decision-making appear to be relatively well understood (see Section 2.2). Decision research has clearly come a long way from the early perspectives of optimal decision-making under risk to concurrent sophisticated models of human decision-making in complex task environments. We now know that both intuition and deliberation play a crucial role in human cognition. In fact, examples have accumulated regarding where and how intuition guides complex decisions in organizations (Akinci & Sadler-Smith, 2012; Dane & Pratt, 2007). Huang and Pearce (2015), for example, found that experienced business angels use sophisticated combinations of deliberate analysis and intuitive judgments when comparing funding proposals of start-up firms. Accordingly, having established the importance of both intuition and deliberation, research appears to shift to some degree toward examining what Dane and Pratt (2007) termed the "use factors" (p. 47) of intuition—that is, toward establishing sound knowledge regarding when decision-makers actually refer to intuitive versus deliberate decision-making. Currently, this constitutes an important knowledge gap for understanding decision-making in complex task environments: Although we have a rather clear idea of how intuitive and deliberate decision modes work, the conditions under which decision-makers actually utilize those decision modes remain largely unclear. Both naturalistic decision-making, with its heroic model of decision-makers, and judgment and decision-making research, with its neglect of situational contingencies, are not sufficiently informative to answer this important question. In order to better understand human decision-making in complex tasks, we therefore need empirical work based on complex, real-life tasks to help us understand contingencies of the choice of decision modes. The following chapters are meant to contribute to this task.

2.2 TWO TYPES OF INFORMATION PROCESSING AND A CONTINUUM OF DECISION-MAKING MODES

A well-rooted perspective in cognitive psychology is the notion that two different types of cognitive systems govern information processing and decision-making in the human brain (Evans, 2008; Evans & Stanovich, 2013). On the one hand, decision-making can be a fast, automated, and effortless process of information evaluation. On the other hand, it can be slow, conscious, and deliberate. Information processing as well as the underlying neural processing structures differ substantially between these two types of processing systems (Evans, 2010). Type 1 processing is fast and automated and describes the situation in which information is being processed and a decision is being made without or with very limited conscious consideration. Most decisions in our daily lives are of this type, as it enables cognitively effortless responses to routine situations as well as quick responses to threatening situations. Type 1 processing allows quick decision-making and does not tax the limited capacity of working memory in the human brain. Type 1 processing produces responses that are grounded in the information that has been activated in the brain at the time of processing, which can lead to an effective and holistic use of previously learned information that would not be fully available consciously. For example, when driving a car, knowledge about the task of driving is activated; for an experienced driver, the process of driving involves little to no conscious thought and thus little to no conscious effort. However, individuals who have very little driving experience will likely be unable to manage the task with type 1 processing and, instead, will have to contribute much more cognitive effort to decide, for example, when and how to use brakes, lights, and throttle, which subsequently leads to less efficient and safe driving. As type 1 processing occurs primarily unconsciously, decision-makers are often not even aware that they are making a decision in a decision situation and often find it difficult or even impossible to explain the reasons for making a specific decision when asked.

Conscious thinking about information and decision options is type 2 processing. Type 2 processing allows the decision-maker to select which information to include in a decision-making process and evaluate this information in an abstract way—for example, accounting for heterogeneous goals or attributing different weights to different inputs. Type 2 processing also allows for the inclusion of unrelated information and facilitates abstract, controlled, and rule-based responses to a specific situation. Type 2 processing is much slower, allows only serial rather than parallel processing of cues, and is restricted by the limited capacity of working memory. In addition, it is affected by the general intelligence of the decision-maker. This means that while engaging in effortful type 2 processing, decision-makers have only limited capacity to engage in other cognitive tasks. Thus, the application of type 2 processing evokes opportunity costs in the form of an inability to attend to other cognitive tasks (Kurzban, Duckworth, Kable, & Myers, 2013). Type 1 processing is the default option, whereas type 2 reasoning is only activated when a trigger is in place, which might be the case in complex or unfamiliar situations as well as for difficult and abstract problems (Evans, 2010).

The two types of information processing interact in a nonsymmetrical way. Whereas the application of type 2 processing requires conscious intention and purposeful commitment of limited cognitive resources, type 1 processing is an automated process triggered by the perception of cues or a need for a certain decision (Evans & Stanovich, 2013). Upon experiencing a situation that requires information processing or decision-making, the automated type 1 process creates a default response, which a decision-maker may or may not complement or substitute with a type 2 processing response (Evans, 2010). The resulting combination of default automated processing and voluntary deliberate intervention is accordingly known as the default interventionist dual-processing model (Evans, 2010; Evans & Stanovich, 2013). As the degree to which a decision-maker relies on type 1 versus type 2 processing varies between decisions, the two processing types lead to a continuum of possible decision modes, ranging from fully subconscious and quick decision-making involving no use of type 2 processing to extensive and systematic information evaluation based on an extensive use of type 2 processing, which then completely or to a large degree overwrites type 1 solutions. In this book, we will refer to the first archetype as intuitive decision-making and the second archetype as deliberation. Although processing types are innate and distinct cognitive mechanisms, the choice of decision modes depends on the perceived need for reasoning and varies on a continuum (Evans, 2010). The goal of this book is to evaluate important contingencies affecting the choice of decision modes in complex task environments, which are a specific case of a decision situation as defined in Section 1.3.

Decision mode describes the degree to which type 2 processing is used in a specific decision (cf. Evans & Stanovich, 2013). Based on the dual-processing views outlined previously, a decision can be made at any point on the decision mode continuum. A decision-maker can fully rely on his or her intuition; can deliberate somewhat, for example, by integrating outside information through a quick glance at a written note; or can give a decision full and deliberate consideration by attempting to systematically identify the best option for a given situation. The main interest of this book is to understand when decision-makers will choose more intuitive versus more deliberate decision modes contingent upon situational and individual characteristics. The ability to decide when to trust intuition as well as the degree to which one should engage in a deliberate decision mode

---

TWO TYPES OF INFORMATION PROCESSING

Type 1: fast, automated, subconscious
Type 2: slow, deliberate, conscious

PROCESSING TYPES VERSUS DECISION MODES

Processing types: binary, separate cognitive mechanisms
Decision modes: Continuum from intuitive to deliberate; can combine processing types

has been called metacognition (Thompson, Prowse Turner, & Pennycook, 2011). Metacognition monitors type 1 processing and determines the extent to which type 2 processing should be used based on what Thompson and colleagues termed the "feeling of rightness"—that is, the impression that the initial, intuitive decision option should be taken (cf. also Alter, Oppenheimer, Epley, & Eyre, 2007). Thompson and colleagues studied the sociocognitive processes underlying this form of metacognition (Thompson et al., 2013). However, we currently have only a very limited understanding of the internal and external conditions that form and affect this feeling of rightness as well as the cognitive and neural processes on which it is based. Hence, we do not fully understand when decision-makers rely on intuition versus deliberation—in particular, in complex, high-stakes professional work environments. Before we turn to this question, however, it is first necessary to develop a deeper understanding of what intuition and deliberation are as well as which conditions favor one decision mode over the other.

2.3   INTUITION

2.3.1   WHAT IS INTUITION?

Intuition is the decision mode most closely related to type 1 processing. Research on intuitive decision-making has a long tradition in various disciplines (see Akinci & Sadler-Smith, 2012; Hogarth, 2010; Salas, Rosen, & DiazGranados, 2010, for reviews), and a variety of definitions have been proposed. These definitions agree that intuition is a subconscious process that operates largely autonomously (i.e., without a specific conscious intent to use the intuitive decision mode) as well as that intuition does not tax the limited working memory capacities of the human brain. Being a decision mode associated with type 1 processing, intuition is the default response of decision-makers in any given decision situation, which then may or may not be overruled by deliberate thought. In a broad conceptualization, intu-

CORE FEATURES OF INTUITION

- Subconsciousness
- Fast
- Holistic
- Affectively charged

ition encompasses both the cognitive process of fast and holistic information processing ("intuiting") and the resulting decision option ("intuition outcome," cf. Dane & Pratt, 2007). The intuition process builds on holistic associations and affective judgments that are subconsciously referenced when one perceives a decision situation (Dane & Pratt, 2009).

Decision outcomes based on intuition are typically perceived as natural, literally "intuitive" solutions. That is, the decision-maker often has the impression that his or her course of action in a certain decision was the only reasonable way to proceed, which oftentimes leads to the impression that there was no "real" decision at all. If an intuitive decision does not create such a feeling of ease, deliberation is often triggered (Alter et al., 2007). The perception of "rightness" associated with intuitive judgments leads to a feeling of confidence regarding decisions made with an intuitive decision mode (Sinclair, 2005).

Intuition as a decision mode is characterized by four characteristics: Intuition is a *subconscious* mode based largely on type 1 processing. Intuition involves *rapidly produced holistic* associations and produces *affectively charged* judgments (Dane & Pratt, 2007, p. 36). These core characteristics of intuition as a decision mode have various implications for decision-making in complex task environments. We will discuss the most important ones in turn.

2.3.1.1   The Intuition Process Is Subconscious

The intuition process runs in a largely automated and subconscious way. This does not necessarily imply that decision-makers are unaware of the cues they perceive (Newell & Shanks, 2014). However, information processing and the related decision-making process itself remains on a subconscious level. It is only the outcome of the intuition process that can be consciously perceived, for example, as a feeling that a specific decision option should be chosen (Evans, 2010). This perception of rightness (or the lack thereof) is an important input for the application or nonapplication of type 2 processing, as it potentially triggers the impression that deliberate thought is warranted (Alter et al., 2007).

The notion that intuiting is a subconscious process has two main implications. First, relying on intuition, a decision-maker is typically unaware of how he or she reached an intuitive decision, which contrasts with deliberate decision-making, with which the applied decision rules are either cognitively chosen or at least accessible through further deliberation (Betsch & Glöckner, 2010). The social implication of this inability to recall the decision process in retrospect is that decision-makers applying intuition will often be unable to explain a decision after it is made, which might be a substantial problem in many complex task environments. For example, when called to defend a decision in an after-action review or debriefing, which is common practice in many organizations acting in complex task environments, or when dealing with diagnostic errors in medicine, the inability to cognitively reproduce the rules that governed an intuitive decision might be a substantial problem for intuitive decision-makers.

The second implication of the subconscious nature of intuitive decision-making is that intuition is not hampered by the limited capacity in working memory, which has long been noted to greatly limit conscious multitasking and multi-cue processing (Miller, 1956). Intuition is based on type 1 processing, which is capable of processing cues in parallel, thus making it possible to consider a large number of inputs and access many prestored memory patterns simultaneously (Dijksterhuis, 2004). This feature of the intuitive decision mode is particularly useful, since the capacity for subconscious information processing is, in contrast to working memory capacity and analytical capabilities, not related to general conceptions of intelligence (Evans & Stanovich, 2013). Thus, intuitive information processing capacity is, on a ceteris paribus basis, equally effective for highly intelligent as for less intelligent decision-makers. Rather than intelligence, experience plays a key role when determining the effectiveness of intuitive decision-making (cf. Section 2.3.1.2).

2.3.1.2   Intuition Is Fast

Intuitive decision-making is a very fast way to approach a decision situation that results in quick, often almost instantaneous decisions. The speed of intuitive decision-making rests on its link with type 1 processing in the dual-process understanding of human information processing and decision-making. Type 1 processing can handle a large amount of information in a very short time frame. Whereas deliberate thought evaluates cues sequentially (one at a time), intuitive decision-makers process cues in parallel (all at the same time). This allows for the establishment of links between cues (leading to a holistic understanding of situations, cf. Section 2.3.1.3) and strongly accounts for the high speed with which intuitive processes produce decision options in complex task environments (cf. Betsch & Glöckner, 2010). This ability of intuitive decision-making is a substantial asset in complex task environments, in which time pressure is often a critical problem. In general, it helps us navigate the multitude of decisions we must routinely make in our daily lives to the extent that many of our decisions never even attract our cognitive attention (Bargh & Chartrand, 1999).

2.3.1.3   Intuition Is Holistic

Intuition draws only on all pieces of prelearned and perceived information that is activated and salient during the intuition process. Information for which this is not the case is not considered. This implies that intuiting can draw on external cues (e.g., sensory input) only if those are subconsciously perceived as relevant for the given task at hand (Betsch & Glöckner, 2010). Intuitive decisions are thus typically heavily contextualized and situated in the experience and mental frame of the decision-maker. As a result, intuitive decision-makers will arrive at very different decisions based on their individual experience, mind-set, and skill set.

The holistic character of intuition is grounded in a process of pattern matching that links external stimuli to internal, prelearned patterns, information, structures, or, if those are not available, simple decision heuristics to determine the relevance and meaning of perceived cues (Kahneman & Klein, 2009). Pattern-matching can be a powerful tool to overcome complexities in many decision situations as it is a fast way to develop a workable solution to a complex problem—a topic we will discuss in more depth in the following. Because of this pattern-matching feature, intuition can be described as an associative process (Epstein, Pacini, Denes-Raj, & Heier, 1996).

An additional implication of the holistic nature of the intuition process is that a decision-maker cannot willingly exclude information from the intuition process even if it would be reasonable to do so (Betsch & Glöckner, 2010). The upside of this is that, through intuiting, a decision-maker is capable of accessing knowledge he or she could not recall or verbalize deliberately (tacit knowledge). This also implies that an intuitive decision mode cannot draw on abstract thinking, a deliberate application of rules, selective information inputs, or any form of selective information acquisition (Betsch & Glöckner, 2010). Intuition is, in particular, not capable of aligning information that has not been activated beforehand (Simmons & Nelson, 2006). Thus, the decision context is of critical importance for intuitive decision-making.

2.3.1.4 Intuition Produces Affectively Charged Judgments

Affective experiences (moods and emotions) are important drivers of human perceptions, decision-making, and action (Lerner & Keltner, 2000). Many emotions are
short-wired in the human brain, meaning that emotions are triggered more quickly
than conscious perceptions of a certain situation (Lang & Bradley, 2010). Therefore,
emotions are a strong factor regarding decision outcomes of both intuitive and deliberate decision modes (Haidt, 2001). Deliberation allows, to some degree, critical self-
evaluation and thus makes it possible to limit or even negate the effect of emotions
on decision-making if a decision-maker considers such a step necessary. This control
mechanism is weaker in the intuitive mode. If sufficiently salient, emotions might
dominate intuitive decision-making as they often provide an immediate stimulus
for action in a decision situation if this decision situation evokes sufficiently strong
emotions (Kahneman & Frederick, 2002). Perceiving anger, for example, implies
aggressive actions, whereas the decision to flee comes naturally after experiencing
fear (Roseman, 2013). If a decision-maker feels angry during an intuitive decision,
a decision outcome that does not consist of an aggressive action is quite unlikely
when he or she relies solely on intuition. This link between emotion and decision is
not direct, however, as learned coping behaviors moderate the effect of emotions on
decision-making and action (Lazarus, 2006). Soldiers trained to withstand fear in a
battle, for example, are less likely to intuitively react with flight after experiencing
the fear that comes naturally in such a situation compared with individuals who did
not undergo the same training and socialization processes. Still, it is safe to say that
emotional experiences have a stronger effect on the outcome of an intuition process
than on the outcome of a deliberation process. This implies that a decision-maker
applying an intuitive decision mode will often come up with entirely different decisions when entering the decision situation angrily versus fearfully even if all cues
available to him or her outside of the emotional experience are identical.

The affective nature of the intuition process can be either a strength or a weak-
ness. Emotions provide immediate response options to potentially dangerous or critical situations. The ability of the intuitive decision mode to use this input directly
helps to produce very fast ad hoc solutions in critical situations. However, the emotional effect on decision outcomes can also be detrimental to decision quality, as acting on one's affective experience of a situation might often be undesirable in complex
task environments in which decision-makers might, for example, need to overcome
fear in order to perform effectively. This fourth core characteristic of intuition further underlines the fact that the intuitive decision mode cannot be understood as a
means to produce optimal decisions in the sense of a rational choice. Instead, the
relevant question to ask is one of appropriateness. In the following, we will discuss
the strengths and weaknesses of the intuitive decision mode in greater detail.

2.3.1.5 Intuition Has Different Functions

Concluding this general description of intuition, it is important to note that intuition is not merely an input for decision-making and problem solving; it also has at
least two other important functions. First, intuition serves as an input for creativity based on the ability to produce divergent mental associations (e.g., Crossan,

Lane, & White, 1999). Second, intuition is also important for moral judgments, which are typically based on quick, subconscious, and affective pattern-matching processes (Dane & Pratt, 2009; Haidt, 2001). However, the most relevant aspect of intuition for our understanding of human actions in complex task environments is its application as a tool for problem-solving and decision-making in judgmental decision tasks. In the following, we will study the advantages and disadvantages of the intuitive decision mode in greater detail.

FUNCTIONS OF INTUITION

- Creativity
- Moral judgement
- Problem-solving

2.3.2 INTUITION AS A PROBLEM OR AN ASSET

Perspectives on the usefulness of intuition as a judgment and decision-making tool diverge into two broad camps. Although some researchers understand intuition as a mental shortcut prone to biases and systematic miscalculations (e.g., Simmons & Nelson, 2006; Tversky & Kahneman, 1974), others conceptualize it as a sophisticated form of reasoning that is suited to solve problems that are too complex to be solved with deliberate thought (e.g., Betsch & Glöckner, 2010; Nemeth & Klein, 2011). Both perspectives rest on solid empirical evidence. Thus, it comes as little surprise that currently no consistent understanding exists of how effective, and thus advisable, intuition as a decision strategy is. Considering the sound arguments advanced by both camps, it appears very unlikely that such an understanding could actually exist. Instead, a more fine-grained understanding of the advantages and disadvantages of intuition is required to provide recommendations for situations in which intuition is an appropriate decision mode as well as situations in which it is not. In the following, we will briefly outline the controversial debate concerning the usefulness of intuition for approaching the type of decisions that are the focus of this book: judgmental decisions under complexity and at least moderate time pressure.

As outlined in Section 2.1, intuition became the focus of intense scholarly attention in the 1970s. A research stream emerged that takes a decidedly skeptical perspective on intuition as a decision mode, heralded by the work of Amos Tversky and Daniel Kahneman. Advocates of this view, which has been termed the heuristics and biases perspective (e.g., Kahneman & Klein, 2009), define intuition as a mental shortcut that simplifies potentially complex tasks through applying simple heuristics, which, in turn, introduce biases and limit decision quality compared to deliberate reasoning. In numerous experiments, researchers following this perspective unearthed a substantial array of such reproducible biases that reduce decision quality when an intuitive decision mode is applied. The heuristics and biases stream clearly outlines some core deficiencies of the intuitive decision mode.

An exemplary bias is the inability of intuitive decision-making to account for probabilities and base rates. A famous example for a bias of this broader type is the well-known Linda problem (e.g., Epstein, 1994). In this often-replicated experiment, decision-makers receive a description of Linda, who is an outspoken person. She lives alone, has a university degree in philosophy, has participated in antinuclear demonstrations during her studies, and is concerned with issues of social justice. Then, decision-makers are tasked to rate how likely they think it is that Linda is a bank teller, a feminist, or both. In most cases, decision-makers find it intuitively more likely that Linda is both a bank teller and a feminist than that she is just a bank teller, despite the fact that being a bank teller and a feminist is a subgroup of being just a bank teller; that is, the bank teller answer includes the option that she is a feminist as well. This is only one example of a variety of reproducible systematic errors that individuals make when they employ an intuitive decision mode. Another potentially dangerous category of problems that occur frequently in intuitive decision-making is attribute substitution. Attribute substitution describes the tendency of decision-makers using an intuitive mode to substitute attributes of a decision problem such that the problem is simplified. Kahneman and Frederick (2002), for example, asked participants to listen to a job interview and then replaced the question of how good the job candidate would be for the department with the related but different question of how impressive the job interview was. This led to poor candidate selection performance. An example for this class of systematic biases is availability bias, with which the most easily accessible cue in a decision is overvalued against less accessible cues. In complex task environments, these biases represent a substantial problem as it is quite possible that salient cues will be overrated in comparison to less salient ones. Such a pattern has, for example, been shown in medical diagnosis. There, diagnosis errors might be partially attributed to availability biases in intuitive judgments such that symptoms that are easily recognized are weighed much more heavily than symptoms that are more difficult to perceive even if the decision-maker recognizes that both symptoms exist, thus leading to misguided diagnoses (Mamede et al., 2010). In such cases, the complexities inherent in a decision problem, for example a diagnosis for a patient with multiple symptoms, are not solved but rather circumvented with a reasonable but potentially not ideal or even outright wrong decision.

The critical perspective on intuitive decision-making also made an important point regarding when intuitive decision-making is clearly inferior to a deliberate decision mode. In optimization decisions, with which it is possible to arrive at a good solution by applying abstract rules drawn, for example, from statistics or the broader field of mathematics, deliberation will typically perform better provided the decision-maker is cognitively at least broadly aware of the mathematical, statistical, or logical principles and rules that determine optimality in these types of decision situations. In these situations, intuition is a mental shortcut that reduces complexity by referring to a simplified problem or judgmental operation whereas deliberate decision-making can produce an objectively optimal solution to the respective problem (Tversky & Kahneman, 1974). Tversky and Kahneman's conceptualization of intuition as heuristics is relatively unrelated to domain knowledge. In fact, Kahneman and colleagues found that, when applying an entirely intuitive decision mode, even experts are subject to evaluation errors, although they are naturally much quicker to identify and resolve their errors once they deliberate on the decision problem (Kahneman & Frederick, 2002). One important conclusion is that intuition as heuristics is helpful to assess probabilities and predict values in simple judgmental tasks but produces severe decision errors and reproducible biases in more complex environments (Kahneman & Frederick, 2002; Tversky & Kahneman, 1974).

Therefore, intuition as understood in the way of a simple decision heuristic is often inappropriate for complex task environments (Baylor, 2001).

In contrast, other scholars forward a positive perspective of intuition as a decision-making tool (e.g., Betsch & Glöckner, 2010; Dane & Pratt, 2007; Klein, 1995). These positive perspectives on intuition are grounded in the ability of intuitive decision-making to produce fast and holistic solutions to complex problems (see Section 2.3.1) based on the extensive subconscious processing capacities of the human brain, which are not fully employed in deliberate decision-making. The cornerstone of this conceptualization of intuition as a powerful tool to solve highly complex decision problems, and thus the major difference to the critical perspective on intuition, is the central role that experience plays in this conceptualization of intuitive decision-making, which we will discuss in the following section. Despite their generally positive view of intuition as a powerful decision-making tool, the advocates of intuition also acknowledge that intuition is primarily suitable for a specific set of tasks (Kahneman & Klein, 2009). In the following, we will discuss both aspects—experience as well as task structure—in turn.

2.3.3 THE ROLE OF EXPERIENCE IN INTUITIVE DECISION-MAKING

A helpful concept to align the critical and positive perspectives on intuition is the distinction between mature and immature intuition (Baylor, 2001). As we discussed previously, one of the defining features of intuition is that it accesses and processes cues holistically. Still, this feature is only positive for decision outcomes if the decision-maker also has a sound understanding of what these cues mean, both in isolation as well as dependent on each other and the broader environment. Observing smoke on a ship might simply indicate fire for a less experienced decision-maker, whereas a more experienced decision-maker might be able to derive conclusions regarding the position and size of the fire as well as the likely cause and potential dangers, subconsciously aligning the visual observation with other perceived or stored information in the same amount of time.

MATURE VERSUS IMMATURE INTUITION

Immature intuition: based on guesswork or simple schemas
Mature intuition: based on complex expert schemas

For experienced decision-makers, the holistic process of cue identification allows for a pattern-matching algorithm that builds on a quick and almost effortless simulation of decision options based on learned memory patterns (Klein, 1995)—typically called "expert schemas" (Dane & Pratt, 2007).

Expert schemas develop through repeated exposure to a specific type of situation and are thus domain-specific (Evans, 2006; Lieberman, 2000). Expert schemas help decision-makers understand and predict how an observed situation might develop as well as how specific courses of action might potentially affect this situation. Inexperienced decision-makers, those employing immature intuition, lack these pre-stored memory patterns and thus have no ground on which effective pattern-matching triggered by holistic cue processing can be built. Instead, their intuition draws on heavily simplified representations of reality or even mere guesswork, as their inability to relate cues to outcomes based on relevant and rich expert schemas hampers effective pattern matching and, with that, effective intuiting. Even though simplified pattern-matching procedures might produce appropriate solutions in many cases, in complex situations they are prone to fall prey to judgment errors and the severe systematic biases discussed previously (Tversky & Kahneman, 1974). In the case of mature intuition, when automatic pattern-matching processes draw on well-developed expert schemas, the ability of intuition to employ knowledge that is not consciously accessible can be most effective. Mature intuition crucially hinges on domain-relevant knowledge and enables the proficient decision-maker to process a huge amount of information quite quickly. In the best case, the experienced decision-maker would then be able to develop successful intuitive decision options even for novel problems based on prior experience facilitating mental simulations to anticipate and evaluate possible outcomes (Dijkstra et al., 2013; Evans, 2006). In any case, the expert schemas allow for effective mental simulations and pattern matching for familiar problems or problems that resemble well-known ones, thus rapidly producing good intuitive decisions. The effectiveness of intuition is therefore contingent upon the depth and coherence of learned patterns (Betsch & Glöckner, 2010; Dane & Pratt, 2007). Benefitting from its effective type 1 processing, mature intuition aligns all prestored information from memory that is activated by salient contextual cues in the decision environment (Betsch & Glöckner, 2010). Decision-makers who can draw on mature intuition are able to relate salient cues to stimuli and situations that they experienced previously. As a result, tacit knowledge gained in prior decision situations can be applied to a specific decision situation.

Thus, complex and domain-relevant expert schemas enable experts to make accurate intuitive judgments even in complex tasks (e.g., Chase & Simon, 1973; Dane, 2010). Expert schemas contain knowledge about an issue or a certain stimulus, related attributes, and the connections between these attributes (Dane, 2010; Fiske & Taylor, 2013). Still, the effectiveness of expert schemas hinges on the domain of expertise. Chess masters' highly complex cognitive schemas help them to easily recall the placement of chess pieces on a board (Chase & Simon, 1973) but do not help solve decision problems in other fields of expertise. Experience is, as far as intuitive decision-making is concerned, highly contextual; and so are expert schemas. Expert schemas differ between novices and experts regarding extent and complexity. Experts have both more schemas available (i.e., they have developed schemas for a broader range of situations) as well as deeper and more complex schemas, allowing for more effective and accurate intuitive decision-making (Dane, 2010). Furthermore, experts organize knowledge in a way that makes it more accessible, efficient, and functional in comparison to novices. Expert schemas also display more and stronger interrelations between different concepts than the schemas of novices (Bédard & Chi, 1992). In sum, experts develop the capacity to effectively and efficiently link cues they perceive consciously or subconsciously with other cues and possible courses of action and are thus able to make intuitive decisions much more effectively than novices.

EXPERT SCHEMAS

Knowledge is structured in contextualized schemas. Expert schemas are complex and have a high degree of connectedness and coherence, allowing for effective cross-referencing and mental simulation.

Based on research on real-life decision tasks and the observation that decision-makers often perform very well when relying on intuitive decision-making in complex task environments, the research stream known as "naturalistic decision-making" (Lipshitz et al., 2001b) placed particular emphasis on understanding the link between experts' cognitive schemas and intuitive decision-making in complex task environments with judgmental decision problems. The naturalistic decision-making stream developed a variety of models that aim to explain how expert decision-makers make such effective use of the intuitive decision mode. The recognition-primed decision model is the most relevant of the models developed in this empirically driven research stream. Based on a series of field studies and interviews with expert decision-makers who made decisions under extreme time pressure (Kaempf, Klein, Thordsen, & Wolf, 1996; Klein, Wolf, Militello, & Zsambok, 1995; Klein, 1995), the recognition-primed decision model conceptualizes the process of intuitive decision-making in complex tasks as a combination of situation assessment and mental simulation. Experts assess a situation to generate a plausible course of action and evaluate this course of action via mental simulation (Klein, 1995). In a decision situation, a decision-maker evaluates the situation (recognition stage) and then reacts based on the first decision that comes to mind. Although a form of intuitive heuristic in the sense of Tversky and Kahneman (1974), this form of decision-making is effective if decision-makers can develop the initial solution based on subconscious mental modeling and pattern matching, drawing on prestored relevant expert schemas—that is, if they can apply mature intuition. Decision-makers with relevant task experience are often able to do so (Klein et al., 1995). Experience allows the decision-maker to categorize situations, perceive typical situation characteristics, and, as a consequence, arrive at appropriate decisions. Furthermore, experience enables the decision-maker to mentally simulate actions to detect unintended, unacceptable consequences. Simple decisions can be solved by pattern matching alone (a purely intuitive decision mode). However, if a decision is complex or new, conscious mental simulations become increasingly important (Kaempf et al., 1996). This form of recognition-primed decision-making is therefore not fully intuitive but, rather, involves conscious elements—that is, a property of type 2 processing (Evans & Stanovich, 2013). In a continuous scale ranging from full intuition to full deliberation, recognition-primed decision-making falls on the intuitive side but not fully so. The naturalistic decision-making stream thus added insights to our understanding of how experience relates to effective intuitive decision-making in complex task environments based on illustrative empirical fieldwork. Yet, what we did not learn from naturalistic decision-making is the degree to which decision-makers actually apply pattern-matching and mental modeling as well as whether that degree is contingent upon situational characteristics. With its specific focus on decision-making under high time pressure, naturalistic decision-making is ill equipped to answer this question, as it focuses on a decision situation in which the application of deliberation is generally impossible.

In sum, it seems clear that experience is a prime contingency condition for the usefulness of intuitive decision-making. The main benefit of intuition is its speed and the limited use of cognitive capacity. Although novices can also use intuition in a subconscious and quick manner, they lack the necessary cognitive schemas to benefit from the holistic nature of intuitive decision-making. Experts, in turn, benefit from the holistic nature of the intuition process, as they have the ability to make full use of the broad information processing capacities and abilities to subconsciously access tacit knowledge. This does not imply that intuition is always the appropriate choice for expert decision-makers nor that intuition is always a poor choice for novices (cf. Gigerenzer, 2008). Still, the conclusion that intuitive decision-making is better suited for experts than for novices is certainly warranted. Still, experience is generally a positive force in decision-making, as it also substantially benefits deliberate decision modes, which we will discuss later. For a full understanding of the benefits and downsides of intuitive decision-making, we must also consider another contingency condition: the task structure.

2.3.4 THE ROLE OF TASK STRUCTURE IN INTUITIVE DECISION-MAKING

Intuitive decision-making is not equally well suited for all types of decision tasks. As discussed earlier, intuition is inferior for optimization tasks as well as in all situations in which mathematical calculations are important. However, in the field of judgmental tasks under varying degrees of complexity, the type of decision situation we are interested in in this book, there are situations that are better or worse suited for intuitive decision-making.

When considering the effects of task structure on the effectiveness of intuitive decision-making, the relationship between task cues and possible outcomes is of central importance. In particular, it is critical whether the link between cues and outcome is stable over time—that is, whether a specific cue reliably relates to a specific outcome, whether the relationship differs over time, and whether most of the cues are coherent; that is, cues observed in a specific situation are not contradictory. Another important aspect of the task is its novelty. As established previously, intuition benefits from field experience; thus, whether a decision-maker had time to develop experience as well as whether a task is new is therefore important.

The first critical aspect of task structure concerns the stability of the link between cues and outcomes (Kahneman & Klein, 2009). As discussed earlier, in order to be effective, intuition requires learned expert schemas relevant for the specific situation. Expert schemas must be learned. Still, learning requires that the relationship between cause and effect can be understood (at least on a subconscious level) and that such relationships are relatively constant between different occurrences of the respective scenario. If this is not the case, expert schemas can either not be developed or are, in the worst case, misleading. In particular, management research is rife with examples in which causal ambiguity, that is, unclear relationships between cause and effect, lead to negative learning curves (e.g., Zollo, 2009). This problem has also been discussed in other fields (Hogarth, 2005). If a specific cue indicates outcome A in one case and outcome B in another, it cannot serve as a basis for the development of reliable expert schemas. Worse, if the experience of a decision-maker

---

TASK STRUCTURES FAVORING INTUITION

- Cues are consistent, meaning that most cues indicate the same best decision.
- Link between cues and outcomes is stable over time.
- Task is not too different from previously experienced tasks.

---

rests only on situations in which the cue led to outcome A, his or her intuition will be wrong when type B situations are encountered.

In addition and relatedly, building expertise takes a considerable amount of time and practical exposure, and tasks that do not allow the necessary exposure are also less well suited for intuitive decision-making (Kahneman & Klein, 2009). This refers in particular to very rare or new occurrences. In these cases, however, intuition might still have its use, as intuition also helps to develop novel solutions. Still, the ability to effectively apply subconscious pattern matching is substantially hampered in these cases.

The coherence of cues is also important. Due to its holistic nature, applying an intuitive decision mode means that a decision-maker perceives all cues he or she deems relevant on a subconscious level simultaneousy. Intuitive decision-making does not allow selective cue processing or cue weighting. This is a problem in situations in which cues are inconsistent, for example, when some cues suggest outcome A and others suggest outcome B. In these cases, intuitive decision-making often fails to produce a decision (Betsch & Glöckner, 2010), sparking the need to apply more deliberation (Alter et al., 2007; Thompson et al., 2013).

Intuitive decision-making is also not suitable if a decision requires the inclusion of perspectives of different persons and/or accounting for inconsistent goals as well as, as discussed previously, whether decision-makers must justify their decisions to others, since the rules on which an intuitive decision is based cannot be consciously recalled. This is because intuitive decision-making always builds solely on the perspective of the intuitive decision-maker without an ability to include outside information or weigh different perspectives against each other (Lipshitz et al., 2001b).

In summary, we know that intuition as a decision mode rests strongly but not necessarily exclusively on type 1 processing and subconsciously produces fast, holistic, and affectively charged judgments. Intuition is subject to various reproducible biases but can also be a powerful asset for decision-makers because of its speed and ability to relate tacit knowledge to a specific decision problem. The latter advantage requires field-specific experience to fully unfold its potential. In order to develop this type of expertise, an environment that provides learning opportunities as well as stable and visible links between cues and outcomes is required. The effectiveness of intuition as a decision mode is therefore contingent upon situational and intrapersonal characteristics. Having established this core understanding of the intuitive decision mode, we will now turn to the deliberate mode.

2.4 DELIBERATION

2.4.1 WHAT IS DELIBERATION?

Deliberate decision-making is an analytic, controlled process of thinking about a decision that a decision-maker can apply if he or she does not want to follow his or her intuitive decision mode. Various triggers can induce deliberation, in particular a lack of the feeling of "rightness" or fluency that often

CORE FEATURES OF DELIBERATION

- Conscious
- Slow
- Selective
- Analytical

comes with intuitive decision-making (Alter et al., 2007). Deliberate decision-making involves *conscious and rule-based* reasoning and tends to be *slow and taxing on working memory* (Evans & Stanovich, 2013). Although deliberate decision-making is cognitively different from intuitive decision-making, resting heavily on type 2 processing, deliberation still *builds on type 1 processing*, which is faster and, therefore, typically precedes deliberation (Evans, 2008). This implies that decisions made with the deliberate decision mode are not independent of preceding type 1 processing (cf. Section 2.2). We will discuss the core characteristics of the deliberate decision mode in the following.

2.4.1.1   Deliberation Is Conscious and Rule-Based

Decision-makers applying a deliberate decision mode are typically aware of the cues they include in a decision-making process as well as which decision rules they use. As a result, they are much more in control of the decision process than intuitive decision-makers (Evans, 2008). In deliberate decision-making, cues are considered in isolation—that is, as independent pieces of information. This implies that a deliberate decision-maker can select or deselect specific information for the decision process and can consider each cue in its own right. However, the decision-maker must not necessarily do so. One common form of deliberation is the more or less purposeful selection of a decision heuristic, which is then applied to arrive at a certain decision (Betsch & Held, 2012; Gigerenzer & Gaissmaier, 2011). An example of a commonly applied heuristic is the recognition heuristic, in which a decision-maker, facing different options, chooses the option that appears most familiar (Gigerenzer, 2008). For example, with three vessels in distress that all request assistance at the same time, a decision-maker applying this heuristic might consciously choose to prioritize that vessel with which he or she is most familiar, ignoring other pieces of information, thus simplifying the decision situation. Heuristics are deliberate in that the decision-maker chooses which decision rule to apply, but they often still carry strong elements of type 1 reasoning. The recognition mechanism, for example, on which the recognition heuristic is based, is a property of type 1 reasoning.

A decision-maker applying the deliberate mode can also decide to put different weights on cues, valuing some pieces of information higher than other pieces. This selective inclusion and weighting of information allows analytical processing that goes far beyond the abilities of an intuitive situation evaluation, which is grounded in holistic cue processing. Holistic processing, in contrast, means that cues are not considered in isolation but as part of a general (holistic) impression of a specific situation (Hogarth, 2005). Selection and weighing of cues is therefore not possible in the intuitive mode.

Deliberation is also potentially superior regarding the selection of cues. Intuition is restricted to decision-making based on preactivated expert schemas and suffers from related perception filters (cues that do not fit into a preactivated cognitive pattern are potentially ignored), whereas the deliberate mode does not have these restrictions. Thus, deliberation allows for different degrees of general and abstract analyses and a selection of decision options based on clearly and consciously defined rules. Information can be selected based on considerations of relevance; and discrete weights can be placed on each piece of information.

In addition, deliberation also allows for the conscious weighting of goals, prioritizing one goal over the other, which is particularly relevant when goals are conflicting.

A decision-maker might, for example, consider different goals that can be pursued in a given situation (e.g., balance efficiency vs. economy or diverging goals of different parties) and weigh those goals against each other. Deliberate decision-making also allows one to include information that does not reside in the decision-maker's mind, which can also enrich the informational basis on which deliberate decision-making is based. Thus, a decision-maker applying a deliberate mode might come to the conclusion that the information he or she has regarding a decision situation is not sufficiently complete to allow a good decision and thus may look for additional information to complement his or her initial perception, which does not typically happen in an intuitive decision mode.

In addition, a deliberate decision-maker can also select the decision rule that he or she considers most appropriate for a given situation and, in particular, recall which rule he or she used after the decision has been made. This is particularly beneficial in a situation in which decision-makers are under legal or social pressure to defend decisions after they have been made, such as in the case of medical diagnosis or in settings in which after-action reviews are part of the work routine. If decision-makers are able to clearly state why they made a specific situation, interindividual learning is facilitated, and the effectiveness of after-action reviews or comparable techniques as tools to increase group performance increases (Salas, Wilson, Burke, & Wightman, 2006).

However, the necessity of selecting the rule for a decision also has its downsides. Although the choice of the correct rule or processing algorithm might be easy in simple problems, finding an appropriate decision rule will be increasingly difficult if problem complexity increases (Hogarth, 2005). In cases in which decision-makers face many cues and complex goal structures or in which the link between cue and outcome is ambiguous or unknown to the decision-maker, the choice of an appropriate decision rule might be very difficult or even impossible. Examples are complex math problems that cannot be solved if the calculation algorithms are unknown or multi-cue decisions for which decision-makers are overwhelmed by the amount of information they would have to process. The latter problem relates to the slow speed and high cognitive workload that conscious and rule-based processes imply. Heuristics are simple rules that might help overcome these problems in some instances, if a decision-maker decides to use only a limited degree of deliberation or purposefully ignore a substantial amount of the available information. Unlike intuition, deliberate decisions are always associated with a trade-off between decision speed and decision accuracy, as heuristics always use only a small part of the information available for a current situation and imply the use of very simple and partially imprecise decision rules, comparable to the attribute substitution problem of intuitive decision-making as described previously (Kahneman & Frederick, 2002).

2.4.1.2 Deliberation Is Slow and Places Strain on Working Memory

Deliberate decision-making rests on type 2 processing, which loads heavily on working memory capacity (Evans, 2008). The concept of working memory describes those cognitive systems that account for short-term memory capacities as well as conscious data processing and which are thus critical for analytical reasoning and comprehension (Baddeley, 1992). The capacity of the working memory is limited to approximately seven items at a specific time (Miller, 1956). In addition, working memory processes information sequentially, meaning that cues enter the

decision process simultaneously (Betsch & Glöckner, 2010). This causes deliberate decision-making to be a much slower process than intuitive decision-making,
which rests to a larger degree on fast type 1 processing. The capacity to effectively
process cues in a deliberate way, apply abstract rules and algorithms, and arrive at
sufficiently complex decisions that make full use of the advantages of the deliberate
mode is related to traditional conceptions of intelligence (Evans, 2008)—that is, not
everyone is equally well suited to develop sophisticated decisions based on deliberation in a complex task environment.

In addition, applying conscious cognitive capacities is an effortful process, which
decision-makers often experience as taxing and exhausting. Thus, decision-makers
tend to avoid tasks (information search as well as cognitive processing) that place
heavy strain on the working memory or, at the very least, consider other uses for
these limited capacities (Hardy, 1982; Kurzban et al., 2013). The ability to successfully employ working memory is also related to concepts of general intelligence
(Evans, 2008), meaning that individual differences regarding cognitive capabilities
strongly affect the decision quality that a decision-maker relying on a deliberate
decision mode can achieve.

The result is that the capacity of the working memory is a critical bottleneck for
human cognition and decision-making. Deliberate processing is slow and implies
opportunity costs, meaning that a decision-maker cannot commit his or her resources
elsewhere while focusing deliberate thought on a specific decision. Reducing the
cognitive load, for example, by applying simple decision heuristics, opens this bottleneck wider but also results in a loss in capacity for analytical reasoning as well as
a loss of information to be included in the decision. This implies that deliberation
cannot be used in cases of extreme time pressure when instantaneous decisions are
required. Even if time pressure is not prohibitively high, it is typically preferable to
arrive at a decision quickly instead of slowly. The slow speed and high cognitive
workload are therefore substantial disadvantages of the deliberate decision mode for
many complex task environments.

2.4.1.3 Deliberation Is Influenced by Type 1 Processing

As intuition is an automated process, it is not possible to inhibit subconscious information processing. Therefore, the first impression of a decision or problem-solving
task is always subconscious type 1 processing. In fact, type 1 processing is a crucial
input for deliberate decision-making, as it provides the deliberate mode with a continuous influx of cues on which to base decisions (Evans, 2008). This initial assessment and first courses of action gained via subconscious information processing may
be rejected or accepted and thus may or may not trigger deliberate decision-making.
When the deliberate decision mode is triggered, the intuitive decision-making process
has already commenced and serves as an input for deliberation.

However, the initial type 1 processing—that is, intuitive evaluations of a decision
situation—might create a bias for the following deliberations. In general, decision-makers
tend to be inclined to search selectively for information that supports an initial decision;
and information that would cast doubt on the initial decision tends to be ignored. This
bias, which has been called the *einstellung* or "set" effect (Bilalić, McLeod, & Gobet,
2008) is particularly prominent when a decision task offers familiar, salient features

that seemingly relate to a specific decision. In these cases, less salient cues tend to be overlooked, and the deliberation process focuses largely on developing an explanation or justification for the decision already made with the intuitive process. This is a problem, as the neglect of nonsalient or otherwise not immediately evident cues is one of the substantial downsides of type 1 processing. In these cases, deliberation bears the downsides of type 2 processing (slow speed, cognitively demanding, dependent on intelligence) but cannot utilize the advantages of processing cues that the intuitive process would have overlooked (Wilson & Schooler, 1991). This application of deliberation as a justification for intuitive decision-making might be useful in some cases—for example, to avoid the justification problem inherent in intuitive decision-making. However, it is often a reason for concern regarding the effectiveness of deliberate decision-making.

The interaction between type 1 processing and type 2 processing has both positive and negative aspects for decision-making in complex task environments. The discussion underlines the fact that the distinction between intuitive and deliberate decision modes is not as clear-cut as the distinction between type 1 and type 2 processing. Instead of being distinct and separate systems, intuition and deliberation interact. Deliberation typically involves some degree of type 1 processing; and most nonroutine tasks will include some elements of deliberate thought. This has been empirically shown, for example, in the well-known recognition-primed decision model discussed in Section 2.3.3. The high demands regarding cognitive capacities, on the other hand, encourage decision-makers to apply at least some intuition in many decisions (cf. Bargh & Chartrand, 1999). This is particularly true in complex task environments in which time restraints often place pressure on decision-makers to come to decisions relatively quickly.

2.4.2 Deliberation as a Problem or an Asset

Comparable to intuition, deliberation has a variety of advantages and disadvantages that can turn this decision mode into an asset as well as a problem for decision-making in complex task environments. In general, an important advantage of deliberate decision-making is that such processes are more robust regarding decision biases than intuitive decisions. Many reproducible errors made by intuitive decision-makers, such as failure to account for regression to the mean or attribute substitution (cf. Tversky & Kahneman, 2000), can be avoided to some degree if a decision-maker places deliberate thought into a decision, provided that the deliberate process rests on appropriate decision rules. As deliberate decision-making is grounded in rules that are consciously selected, it is much easier to explain and understand decisions after they have been made compared to decisions made with an intuitive mode. Deliberation involves the sequential and potentially selective evaluation of cues (again, if grounded in a comprehensive analytical approach and not in simple heuristics) and can include outside information. Another important advantage of deliberate decision-making is the ability to better adapt to new, unknown situations. As effective deliberation does not (solely) rest on the accuracy of prestored memory patterns (expert schemas), it is a superior strategy for tasks that require the development of entirely new solutions or for which existing expert schemas are misleading (Betsch & Held, 2012).

Although not subject to the biases associated with intuitive decision-making, the deliberate decision-making process is also prone to errors. In order to reach a good decision, three steps need to be completed successfully. First, there is a need to identify the relevant cues. As deliberation does not build on a holistic impression of a situation, selecting and weighing relevant (and only relevant) cues is often difficult. Second, the cues need to be captured in an appropriate way in order to be accessible for deliberate cognitive processing. Finally, appropriate rules for the aggregation of information of the previous steps need to be chosen (Hogarth, 2005). If all of these variables, measures, and rules are applied correctly, the deliberate decision-maker still requires the necessary time, cognitive capacity, and competence to successfully align cues and rules to develop an appropriate decision. Under conditions of uncertainty and environmental complexity, cue selection, measurement, and rule application become increasingly difficult. Consequently, the deliberate decision process tends to be more time-consuming and more likely to produce inappropriate decisions as complexity increases. Because of the high cognitive effort and limited cognitive capacity, the number of cues that can be processed at the same time is small. Deliberation is therefore prone to errors grounded in incorrect rules, incorrect rule application, false or incorrectly weighed cues, and incomplete information.

To develop a full understanding of the conditions under which deliberate decision-making is a good decision strategy, it is again important to look at two key contingency conditions: The role of the decision-maker's experience and the task structure.

2.4.3 THE ROLE OF EXPERIENCE IN DELIBERATE DECISION-MAKING

Experience is beneficial for both deliberate and intuitive decision-making, yet the benefits that experience provides are somewhat different. In intuition, experience is primarily important as the effectiveness of intuitive decision-making hinges critically on the expert schemas that a decision-maker developed through prior exposure to relevant situations or other means of learning. In deliberate decision-making, the learned knowledge has a somewhat different effect. Experienced decision-makers have a clearer understanding of the rules they can and should apply to a specific situation. When applying a heuristic approach as a low-level deliberation strategy, experienced decision-makers know better than less experienced decision-makers when specific heuristics are effective in a given situation (Gigerenzer, 2008). They also find it easier to identify important cues among the variety of cues perceived at any given moment. In consequence, they can apply deliberate decision-making with more certainty and fewer errors than less experienced decision-makers (Dijkstra et al., 2013). In addition, experts have a better understanding of which disassociated knowledge might be important for a specific situation and are therefore better able to determine information requirements for a specific decision (Mamede et al., 2010). Another benefit of experience lies in cognitive processing: Experience allows the processing of larger perceptual chunks. This means experts can process more related information at the same time, making the bottleneck of cognitive processing and working memory load wider for experts than for novices, a mechanism that is grounded in pattern recognition (Chase & Simon, 1973). Relatedly, experts are also better at associating perceived patterns with effective actions and strategies. They recognize bad options more quickly and are therefore better able to focus their cognitive capacity on promising decision alternatives (Raab & Johnson, 2007). Experience is therefore a way to circumvent or at least reduce the problem of the high cognitive workload of deliberate decision-making as experienced decision-makers are able to use their limited cognitive capacities more economically. These effects are related to the ability of experienced decision-makers to categorize decision problems more effectively (Chi, Feltovich, & Glaser, 1981) due to their ability to recognize previously experienced structures and patterns (Pretz, 2008). This effect is comparable to the role that experience plays in intuitive decision-making.

Experience is also helpful for emotion control. As discussed in Section 2.4.1, type 1 processing, which is heavily affected by affective experiences, influences deliberate decision-making. Experience helps decision-makers to understand and potentially control this interlink between type 1 processing and deliberation such that emotions do not create an unwanted bias in deliberate decisions (Fenton-O'Creevy, Soane, Nicholson, & Willman, 2011). Thinking deliberately about a decision, an experienced decision-maker might, for example, discover that his or her initial solution (i.e., the result of automated type 1 processing) was largely guided by a specific emotion and thus take a step back and reconsider the situation. Deliberation can thus create a safeguard against detrimental effects of affect on decision outcomes if a decision-maker is experienced enough to recognize the effect that his or her emotions have on decision outcomes. Finally, experience might also have a negative effect on the performance of the deliberate decision mode as experienced decision-makers tend to lose flexibility regarding the development of new solutions. Through this mechanism, experience can lead to a lock-in effect, with which experienced decision-makers are no longer able to fully exploit the strength of the deliberate mode to consciously develop new solutions (Dane, 2010).

BENEFITS OF EXPERIENCE FOR DELIBERATION

- Broadens repertoire of rules
- Improves rules selection and better understanding of information demand
- Helps process larger chunks of information
- Supports emotion control

Although experience is not as central a contingency condition for deliberate decision-making compared to intuitive decision-making, it is clear that deliberation strongly benefits from experience. Experience helps to economize the use of limited cognitive resources and implies deeper knowledge regarding which rules are available and applicable to a decision situation. Potential lock-in effects of experienced decision-makers are a concern, as they hamper the ability of experienced decision-makers to develop (radically) new solutions.

2.4.4 THE ROLE OF TASK STRUCTURE IN DELIBERATE DECISION-MAKING

In situations in which a best solution exists, deliberation is almost always the best choice (Moxley, Ericsson, Charness, & Krampe, 2012). However, decisions in complex task environments are often not of that kind. Instead, under conditions of uncertainty and complexity, deliberation is a possible but not necessarily the best strategy. Deliberation is particularly suited for decision problems for which a limited number of cues are available that display relatively clear-cut cue relationships with the outcomes. That is, for problems that in some way resemble optimization problems. Deliberation is also well suited for situations in which only relatively few cues out of a potentially large population of cues carry strong predictive power. In both cases, the ability to purposefully select cues helps optimize decision outcomes, whereas the bottleneck problem imposed by the limited working memory capacity is less of a problem under conditions under which only relatively few cues are required to make good decisions. With an increase in the number of cues that need to be perceived and comprehended quickly, the associative and holistic aspects of intuition become increasingly important, in particular if the link of the cues to the outcome is not clearly visible, that is, if tacit knowledge is required to link cues to possible outcomes (Dane & Pratt, 2007). Still, finding the right rule, a precondition for successful deliberate decision-making, is more difficult in tasks with many or very few cues, as cues often signal which rules a decision-maker should or should not use (Evans, Clibbens, Cattani, Harris, & Dennis, 2003). Time pressure is also obviously a critical determinant of a task structure that affects the usefulness of deliberation. As deliberation is costly regarding time and mental capacity, deliberation is not well suited for decision problems under high time pressure (Klein, 1995).

Deliberation is the better choice in situations in which salient cues, which have a strong effect on intuitive decision-making, tend to be misleading or irrelevant for a decision situation (for example, in complex medical diagnosis problems; cf. Mamede et al., 2010). In these cases, recognition-primed intuition will be substantially inferior to deliberation, especially for experienced decision-makers. The same is true for low-level deliberation strategies—that is, the application of simple heuristics. In these cases deliberate decision-makers can consciously ignore or down-weigh the salient (but misleading or irrelevant) cues in favor of less salient cues. Finally, the ability to account for heterogeneous goal structures makes deliberation particularly suited for any decision problems that require compromising or weighing the importance of different goals.

TASK CONTEXTS FAVORING DELIBERATION

- Relatively few cues carry strong predictive power
- Low time pressure
- Salient cues are misleading
- Inconsistent goal structures

In sum, deliberation can be used for a wide variety of decision tasks. The only task characteristic that is clearly prohibitive for a highly deliberate mode is high time pressure. For all other conditions, deliberation is a feasible, yet not always ideal, alternative.

2.5 WHICH DECISION MODES DO DECISION-MAKERS USE: CURRENT STATE OF KNOWLEDGE

Having discussed the two archetypical decision modes in some detail, we will now turn to the core question of our book: When do decision-makers prefer one decision mode over the other in complex task environments? It is relatively clear that the two contingency conditions, experience, and task structure—in particular regarding task complexity—should affect the suitability of decision modes. If we align these considerations with actual decision-making behavior, we are able to derive conclusions regarding the efficiency of decision-making in complex task environments. If we want to understand decision-making and decision outcomes in complex task environments, this is a core question that needs to be answered. Although there has been research on when decision-makers should ideally apply deliberation or intuition, insights into when they actually use a particular mode are scarce—especially in terms of real-life, complex decision tasks. Based on the default interventionist dual-processing model developed earlier, we know that intuition is a default solution. This is also grounded in the observation that decision-makers often tend to avoid cognitively demanding activities (Bargh & Chartrand, 1999). Deliberation must be triggered. Trigger conditions might be grounded in a decision situation. However, the perception of and response to trigger conditions could be contingent upon intrapersonal variables. In the following, we will outline what we currently know about the choice of decision modes both in general and in complex task environments.

Probably the best-researched contingency condition affecting the choice of decision modes in many decision situations is the *intrapersonal preference for a particular decision-making style*. The preference for a specific decision mode is a stable personality trait (Epstein et al., 1996; Evans, 2010; Evans & Stanovich, 2013). This means that, given the choice, some people prefer to make their decisions with a deliberate mode, based on careful information evaluation and the conscious weighting of different alternatives, and others prefer an intuitive mode with quick and subconscious decisions. These preferences are learned and habituated response patterns that individuals demonstrate when confronted with decision situations (Scott & Bruce, 1995). They prescribe, however, only a tendency to follow a specific decision mode; they do not predetermine decision-making. That is, even a person with a strong tendency for deliberation will use intuition under certain circumstances and vice versa.

Another factor that impacts decision mode selection is *affect*. In general, being in a positive mood leads to more intuitive decision-making; and the same is true for feelings of certainty (Dane & Pratt, 2007). Feelings of certainty are caused by emotions such as anger, happiness, and disgust. In contrast, emotions such as anxiety and hope trigger feelings of uncertainty and, subsequently, a tendency for intuitive decision-making (cf. Tiedens & Linton, 2001). The strength of an emotional experience might also have an impact on decision modes. Moderate levels of fear accompanied by little previous experience in a situation foster deliberate decision-making, whereas intense fear, especially when combined with a lack of experience regarding the focal situation, induces (immature) intuitive decision-making (Coget et al., 2011).

The most obvious situational contingency affecting the choice of decision modes is the general *task environment*. In particular, the accessibility of information affects the choice of decision mode. If information is easily accessible and salient cues are easily perceived, the application of intuitive decision-making is likely. In contrast, a lack of salient cues acts as a trigger for deliberation (Söllner, Bröder, & Hilbig, 2013). Additionally, the

> Decision-makers apply rational information processing when facing difficult tasks and difficult-to-access information.
> Rationality is fostered by negative affect and uncertainty-associated emotions.

perceived difficulty of tasks influences the employed information processing mode such that an increased perceived difficulty tends to trigger deliberation (Alter et al., 2007).

Intuitive information processing is more likely to be applied if relevant information is accessible and easy to understand.

Intuition is facilitated by positive moods and certainty-associated emotions.

These insights rest on laboratory experiments with modified reading difficulty; thus, the predictive validity for complex, real-life task environments is unclear. Other authors speculate that perceived novelty as well as motivation might increase the tendency to apply deliberation (Evans & Stanovich, 2013). A related point has been made by the naturalistic decision-making research stream, which found that under high time pressure experts often use intuition although deliberation plays a role in some elements of the decision models of naturalistic decision-making as well (Klein, 1995). In sum, insights into when decision-makers actually use intuition or deliberation in complex, real-life task environments remain scarce.

2.6 SUMMARY

At the conclusion of Chapter 2, it is time to wrap up our discussion of the concepts that are critical for the following parts of the book. In this chapter, we briefly introduced the history of decision-making as an academic discipline, outlining the main insights obtained by some of the most influential streams of knowledge. We also presented the default interventionist dual-processing model of human cognition as the basis of our understanding of information processing and decision-making, establishing that human cognition rests on two interdependent but clearly distinct cognitive processes: an automated, fast, and holistic process (type 1) and a slow, conscious, and deliberate process (type 2). Type 1 processing is the default mode and automatically processes information when a need for a decision is perceived. In contrast, type 2 processing requires a specific trigger to be activated. Intuition and deliberation are the two archetypical decision modes that make use of type 1 and type 2 processing, respectively. Although type 1 processing and type 2 processing are cognitively distinct tools to process information, the decision modes of intuition and deliberation lie on a continuum. Intuition rests heavily on type 1 processing, whereas deliberation is mostly grounded in type 2 processing. Still, decision-makers can gradually apply a specific decision mode—for example, enriching a generally intuitive decision with some degree of conscious mental simulation.

Intuition is fast and holistic (meaning that cues related to a specific situation are perceived as an entirety instead of distinct pieces of information) and draws heavily on pattern-matching based on information activated in the mind of the decision-maker at the time he or she makes the intuitive decision. In the intuitive decision mode, the applied rules and cues included in the decision remain on the subconscious level. Therefore, intuitive decision-makers can typically not recall why they reached a specific decision and which information specifically they used to reach that decision. To be effective, intuition must draw on prelearned insights into specific situations stored in so-called expert schemas. These expert schemas allow pattern matching; that is, they allow the decision-maker to subconsciously relate perceived cues and decision situations to patterns found in comparable, previously experienced situations and thus produce very fast situation evaluations resulting in high-quality decisions.

As intuition is automated, decision-makers cannot abstain from making an intuitive judgment. However, they can override it with some degree of deliberation. Deliberation is a conscious process that allows for the selective inclusion of information and a purposeful selection of decision rules. Deliberation is affected by the preceding intuition process but still allows for generalized and abstract decision-making. Deliberation requires the commitment of limited working memory capacities and thus creates opportunity costs; while a decision-maker deliberates on a decision, he or she cannot do much else. Both intuition and deliberation have their specific strengths and weaknesses, which make them particularly suited for specific situations. Deliberation is the best choice for optimization decisions as well as for decisions in environments in which the link between cues and outcomes is unstable or ambiguous. In these environments, intuition is often not the best strategy, as decision-makers cannot develop the necessary expert schemas on which successful intuitive decision-making rests. Intuition is very well suited for experienced decision-makers in complex but stable environments, in which prelearned expert schemas hold strong predictive value for future decision situations and the inclusion of outside information is not necessary. Still, there is a wide variety of tasks for which both deliberation and intuition are feasible decision strategies, taking into account their specific advantages and disadvantages. Complex task environments, as defined in this book, are among those.

Whereas the conditions for the successful use of intuition and deliberation are relatively well understood, we know relatively little regarding when decision-makers actually apply these decision modes in real-life complex task environments. Answering this question would provide us with a much clearer understanding of decision strategies and decision outcomes in these environments. In the next chapters of this book, we will elaborate on a series of empirical studies in the field of maritime search and rescue (SAR) that aimed to address this important question.

KEY INSIGHTS IN CHAPTER 2

- There are two archetypical decision-making modes: intuitive and rational.
- Intuition can be characterized as quick, associative, automatic, and subconscious, whereas deliberate information processing is described as slow, conscious, and cognitive capacity consuming.
- Intuition incorporates simple heuristics as well as complex associative decision-making processes.
- Situational factors, such as task environment and personal preferences, shape the decision-making process.
- The appropriateness of the decision mode depends primarily on task characteristics and decision-makers' abilities and experience.
- Knowledge regarding which decision mode a decision-maker chooses in a complex task environment is incomplete. Preferences for intuition and deliberation are stable personality constructs. Certain mental states can affect the decision mode. It remains unclear whether and how these conditions shape the choice of decision-modes in real-life complex task environments.
- In the following, we will empirically examine a variety of effects that influence the choice of decision mode in complex task environments.

REFERENCES

Akinci, C., & Sadler-Smith, E. 2012. Intuition in management research: A historical review. *International Journal of Management Reviews*, 14(1): 104–122.

Alter, A. L., Oppenheimer, D. M., Epley, N., & Eyre, R. N. 2007. Overcoming intuition: Metacognitive difficulty activates analytic reasoning. *Journal of Experimental Psychology, General*, 136(4): 569–576.

Baddeley, A. 1992. Working memory. *Science*, 255(5044): 556–559.

Bargh, J. A., & Chartrand, T. L. 1999. The unbearable automaticity of being. *American Psychologist*, 54: 462–479.

Baylor, A. L. 2001. A U-shaped model for the development of intuition by level of expertise. *New Ideas in Psychology*, 19(3): 237–244.

Bédard, J., & Chi, M. T. 1992. Expertise. *Current Directions in Psychological Science*, 1(4): 135–139.

Bernstein, P. L. 1996. *Against the gods: The remarkable story of risk*. New York: John Wiley & Sons.

Betsch, C. 2008. Chronic preferences for intuition and deliberation in decision making: Lessons learned about intuition from an individual differences approach. In H. Plessner, C. Betsch, & T. Betsch (Eds.), *Intuition in judgment and decision making* (pp. 231–248). New York, London: Lawrence Erlbaum Associates, Inc.

Betsch, T., & Glöckner, A. 2010. Intuition in judgment and decision making: Extensive thinking without effort. *Psychological Inquiry*, 21(4): 279–294.

Betsch, T., & Held, C. 2012. Rational decision making: Balancing RUN and JUMP modes of analysis. *Mind & Society*, 11(1): 69–80.

Bilalić, M., McLeod, P., & Gobet, F. 2008. Why good thoughts block better ones: The mechanism of the pernicious einstellung (set) effect. *Cognition*, 108(3): 652–661.

Chase, W. G., & Simon, H. A. 1973. Perception in chess. *Cognitive Psychology*, 4(1): 55–81.

Chen, M., & Bargh, J. A. 1999. Consequences of automatic evaluation: Immediate behavioral predispositions to approach or avoid the stimulus. *Personality and Social Psychology Bulletin*, 25(2): 215–224.

Chi, M. T. H., Feltovich, P. J., & Glaser, R. 1981. Categorization and represenation of physics problems by experts and novices. *Cognitive Science*, 5(2): 121–152.

Coget, J.-F., Haag C., & Donald, E. 2011. Anger and fear in decision-making: The case of film directors on set. *European Management Journal*, 29(6): 476–490.

Crossan, M. M., Lane, H. W., & White, R. E. 1999. An organizational learning framework: From intuition to institution. *Academy of Management Review*, 24(3): 522–537.

Dane, E. 2010. Reconsidering the trade-off between expertise and flexibility: A cognitive entrenchment perspective. *Academy of Management Review*, 35(4): 579–603.

Dane, E., & Pratt, M. G. 2007. Exploring intuition and its role in managerial decision making. *Academy of Management Review*, 32(1): 33–54.

Dane, E., & Pratt, M. G. 2009. Conceptualizing and measuring intuition: A review of recent trends. *International Review of Industrial and Organizational Psychology*, 24: 1–40.

Dijksterhuis, A. 2004. Think different: The merits of unconscious thought in preference development and decision making. *Journal of Personality and Social Psychology*, 87(5): 586–598.

Dijkstra, K. A., van der Pligt, J., & van Kleef, G. A. 2013. Deliberation vs. intuition: Decomposing the role of expertise in judgment and decision making. *Journal of Behavioral Decision Making*, 26(3): 285–294.

Edwards, W. 1961. Behavioral decision theory. *Annual Review of Psychology*, 12: 473–498.

Einhorn, H. J., & Hogarth, R. M. 1981. Behavioral decision theory: Processes of judgment and choice. *Annual Review of Psychology*, 32: 53–88.

Epstein, S. 1994. Integration of the cognitive and the psychodynamic unconscious. *American Psychologist*, 49(8): 709–724.

Epstein, S., Pacini, R., Denes-Raj, V., & Heier, H. 1996. Individual differences in intuitive-experiential and analytical-rational thinking styles. *Journal of Personality and Social Psychology*, 71(2): 390–405.

Evans, J. S. B. T. 2006. The heuristic-analytic theory of reasoning: Extensions and evaluation. *Psychological Bulletin & Review*, 13(3): 378–395.

Evans, J. S. B. T. 2008. Dual-processing accounts of reasoning, judgment, and social cognition. *Annual Review of Psychology*, 59(1): 255–278.

Evans, J. S. B. T. 2010. Intuition and reasoning: A dual-process perspective. *Psychological Inquiry*, 21(4): 313–326.

Evans, J. S. B. T., Clibbens, J., Cattani, A., Harris, A., & Dennis, I. 2003. Explicit and implicit processes in multicue judgment. *Memory & Cognition*, 31(4): 608–618.

Evans, J. S. B. T., & Stanovich, K. E. 2013. Dual-process theories of higher cognition: Advancing the debate. *Perspectives on Psychological Science*, 8(3): 223–241.

Fenton-O'Creevy, M., Soane, E., Nicholson, N., & Willman, P. 2011. Thinking, feeling and deciding: The influence of emotion on the decision making and performance of traders. *Journal of Organizational Behavior*, 32(8): 1044–1061.

Fiske, S. T., & Taylor, S. E. 2013. *Social Cognition* (2nd ed.). London: Sage.

Gigerenzer, G. 2008. Why heuristics work. *Perspectives on Psychological Science*, 3(1): 20–29.

Gigerenzer, G., & Gaissmaier, W. 2011. Heuristic decision making. *Annual Review of Psychology*, 62: 451–482.

Gore, J., Banks, A., Millward, L., & Kyriakidou, O. 2006. Naturalistic decision making and organizations: Reviewing pragmatic science. *Organization Studies*, 27(7): 925–942.

Haidt, J. 2001. The emotional dog and its rational tail: A social intuitionist approach to moral judgment. *Psychological Review*, 108(4): 814–834.

Hardy, A. P. 1982. The selection of channels when seeking information: Cost/benefit vs least-effort. *Information Processing and Management*, 18(6): 289–293.

Hogarth, R. M. 2005. Deciding analytically or trusting your intuition? The advantages and disadvantages of analytic and intuitive thought. In T. Betsch & S. Haberstroh (Eds.), *The Routines of Decision Making* (pp. 67–83). Mahwah, NJ: Lawrence Erlbaum Associates, Inc.

Hogarth, R. M. 2010. Intuition: A challenge for psychological research on decision making. *Psychological Inquiry*, 21(4): 338–353.

Huang, L., & Pearce, J. L. 2015. Managing the unknowable: The effectiveness of early-stage investor gut feel in entrepreneurial investment decisions. *Administrative Science Quarterly*, 60(4): 634–670.

Kaempf, G. L., Klein, G., Thordsen, M. L., & Wolf, S. 1996. Decision making in complex naval command-and-control environments. *Human Factors*, 38(2): 220–231.

Kahneman, D., & Frederick, S. 2002. Representativeness revisited: Attribute substitution in intuitive judgment. In T. Gilovich, D. Griffin, & D. Kahneman (Eds.), *Heuristics and biases: The psychology of intuitive judgment* (pp. 49–81). Cambridge University Press.

Kahneman, D., & Klein, G. 2009. Conditions for intuitive expertise: A failure to disagree. *American Psychologist*, 64(6): 515–526.

Klein, G., Wolf, S., Militello, L. G., & Zsambok, C. E. 1995. Characteristics of skilled option generation in chess. *Organizational Behavior and Human Decision Processes*, 62(1): 63–69.

Klein, G. A. 1995. A recognition-primed decision (RPD) model of rapid decision making. In G. A. Klein, J. Orasanu, R. Calderwood, & C. E. Zsambok (Eds.), *Decision making in action. Models and methods* (2nd ed.) (pp. 138–148). Norwood: Ablex Publishing.

Kurzban, R., Duckworth, A., Kable, J. W., & Myers, J. 2013. An opportunity cost model of subjective effort and task performance. *The Behavioral and Brain Sciences*, 36(6): 661–679.

Lang, P. J., & Bradley, M. M. 2010. Emotion and the motivational brain. *Biological Psychology*, 84(3): 437–450.

Lazarus, R. S. 2006. Emotions and interpersonal relationships: Toward a person-centered conceptualization of emotions and coping. *Journal of Personality*, 74(1): 9–46.

Lerner, J. S., & Keltner, D. 2000. Beyond valence: Toward a model of emotion-specific influences on judgment and choice. *Cognition & Emotion*, 14(4): 473–493.

Lieberman, M. D. 2000. Intuition: A social cognitive neuroscience approach. *Psychological Bulletin*, 126(1): 109–137.

Lipshitz, R., Klein, G., Orasanu, J., & Salas, E. 2001a. A welcome dialogue—And the need to continue. *Journal of Behavioral Decision Making*, 14(5): 385–389.

Lipshitz, R., Klein, G. A., Orasanu, J., & Salas, E. 2001b. Taking stock of naturalistic decision making. *Journal of Behavioral Decision Making*, 14(5): 331–352.

Luce, R. D., & Raiffa, H. 1957. *Games and decisions: Introduction and critical survey*. New York: John Wiley & Sons.

Mamede, S., Schmidt, H. G., Rikers, R., Custers, E., Splinter, T., & van Saase, J. 2010. Conscious thought beats deliberation without attention in diagnostic decision-making: At least when you are an expert. *Psychological Research*, 74(6): 586–592.

Miller, G. A. 1956. The magical number seven, plus or minus two: Some limits on our capacity for processing information. *Psychological Review*, 63(2): 81–97.

Moxley, J. H., Ericsson, K. A., Charness, N., & Krampe, R. T. 2012. The role of intuition and deliberative thinking in experts' superior tactical decision-making. *Cognition*, 124(1): 72–78.

Nemeth, C., & Klein, G. 2011. The naturalistic decision making perspective. *Wiley Encyclopedia of Operations Research and Management Science*.

Newell, B. R., & Shanks, D. R. 2014. Unconscious influences on decision making: A critical review. *The Behavioral and Brain Sciences*, 37(1): 1–19.

Orasanu, J., & Connolly, T. 1995. The reinvention of decision making. In G. A. Klein, J. Orasanu, R. Calderwood, & C. E. Zsambok (Eds.), *Decision making in action. Models and methods* (2nd ed.) (pp. 3–20). Norwood: Ablex Publishing.

Pretz, J. E. 2008. Intuition versus analysis: Strategy and experience in complex everyday problem solving. *Memory & Cognition*, 36(3): 554–566.

Puranam, P., Stieglitz, N., Osman, M., & Pillutla, M. M. 2015. Modelling bounded rationality in organizations: Progress and prospects. *The Academy of Management Annals*, 9(1): 337–392.

Raab, M., & Johnson, J. G. 2007. Expertise-based differences in search and option-generation strategies. *Journal of Experimental Psychology: Applied*, 13(3): 158–170.

Roseman, I. J. 2013. Appraisal in the emotion system: Coherence in strategies for coping. *Emotion Review*, 5(2): 141–149.

Salas, E., Rosen, M. A., & DiazGranados, D. 2010. Expertise-based intuition and decision making in organizations. *Journal of Management*, 36(4): 941–973.

Salas, E., Wilson, K. A., Burke, C. S., & Wightman, D. C. 2006. Does crew resource management training work? An update, and extension, and some critical needs. *Human Factors*, 48(2): 392–412.

Scott, S. G., & Bruce, R. A. 1995. Decision-making style: The development and assessment of a new measure. *Educational and Psychological Measurement*, 55(5): 818–831.

Simmons, J. P., & Nelson, L. D. 2006. Intuitive confidence: Choosing between intuitive and non-intuitive alternatives. *Journal of Experimental Psychology. General*, 135(3): 409–428.

Simon, H. A. 1955. A behavioral model of rational choice. *Quarterly Journal of Economics*, 69(1): 99–118.

Sinclair, M. 2005. Intuition: Myth or a decision-making tool? *Management Learning*, 36(3): 353–370.

Söllner, A., Bröder, A., & Hilbig, B. E. 2013. Deliberation vs. automaticity in decision-making: Which presentation format features facilitate automatic decision making? *Judgment and Decision Making*, 8(3): 278–298.

Thompson, V. A., Prowse Turner, J. A., & Pennycook, G. 2011. Intuition, reason, and meta-cognition. *Cognitive Psychology*, 63(3): 107–140.

Thompson, V. A., Turner, J. A. P., Pennycook, G., Ball, L. J., Brack, H., Ophir, Y., & Ackerman, R. 2013. The role of answer fluency and perceptual fluency as metacognitive cues for initiating analytic thinking. *Cognition*, 128(2): 237–251.

Tiedens, L. Z., & Linton, S. 2001. Judgment under emotional certainty and uncertainty: The effects of specific emotions on information processing. *Journal of Personality and Social Psychology*, 81(6): 973–988.

Tversky, A., & Kahneman, D. 1974. Judgment under uncertainty: Heuristics and biases. *Science*, 185: 1124–1131.

Tversky, A., & Kahneman, D. 1979. Prospect theory: An analysis of decision under risk. *Econometrica*, 47(2): 263–292.

Tversky, A., & Kahneman, D. 1981. The framing of decisions and the psychology of choice. *Science*, 211: 453–458.

Tversky, A., & Kahneman, D. 2000. Judgment under uncertainty: Heuristics and biases. In T. Connolly, H. Arkes, & K. R. Hammond (Eds.), *Judgment and decision making: An interdisciplinary reader* (pp. 35–52). Cambridge: Cambridge University Press.

Wilson, T. D., & Schooler, J. W. 1991. Thinking too much: Introspection can reduce the quality of preferences and decisions. *Journal of Personality and Social Psychology*, 60: 181–192.

Zollo, M. 2009. Superstitious learning with rare strategic decisions: Theory and evidence from corporate acquisitions. *Organization Science*, 20(5): 894–908.

3 Decision-Making in Maritime Search and Rescue

In order to understand the domain of maritime search and rescue (SAR), it is of great importance to also examine the history of sea rescue. In doing so, the specifics of this field become more visible and the changes in its systemic framework over time more comprehensible. The following describes the different roots, reasons, and sources contributing to the institutionalization of maritime SAR organizations. We first take a brief look at the last 200 years to reconstruct the establishment of one of the world's leading rescue services and set the scene. Following this, deeper insights into some aspects of decision-making in the MSAR context should be possible.

Our example is the German Maritime SAR Service (DGzRS*)—that is, the rescue association whose service covered the German-speaking area even before Germany as a nation was founded. The DGzRS may not have been the first organization worldwide with the objective to save life at sea. However, it is a good example of the merging of interests behind the foundation of a charitable organization pursuing this purpose—one established without any governmental influence. Like other famous and well-established charity organizations (e.g., the Red Cross), the DGzRS was founded in the middle of the 19th century—in 1865, a year after the international committee of the Red Cross. One of the obvious reasons why many humanitarian organizations were established during that time traces its roots to changing ethical values and social norms. Other aspects include the unique living conditions of the coastal population, general economic and social aspects in Middle and Eastern Europe, and special incentive structures for wealthy people encouraging investments in charities in the main Hanseatic cities near the North Sea and the Baltic Sea. A comprehensive discussion of the history of sea rescue, lifeboat organizations, and maritime SAR would go beyond the scope of this book. Therefore, we will highlight some interesting aspects that are usually not mentioned in this context rather than presenting broad discussions that can be read elsewhere. As researchers on organizational studies, we are neither historians nor legal experts; thus, we cannot provide highly detailed discussions of the historic and legal matters in this context and, instead, focus on aspects relevant to organizational studies research.

---

* The DGzRS acronym stands for *Deutsche Gesellschaft zur Rettung Schiffbrüchiger*, which, loosely translated into English, means German Association for the Rescue of Shipwrecked (Persons).

3.1   SOCIOECONOMIC FRAMEWORK

3.1.1   COASTAL INHABITANTS

In past centuries, living in coastal regions was both a blessing and a curse. The sea provided rough living conditions for the coastal population—both fishermen and peasants. Periodic floods destroyed the harvest; and communities were typically characterized by poverty, blood ties, and winter struggles for survival. Many inhabitants emigrated to larger cities, signed on as crew members on merchant or navy ships, and spent several years abroad as laborers to either support their relatives at home or return wealthier than when they left. Inhabitants' income consisted primarily of the fishermen's catch as well as crops from their own fields. Yet, sometimes they were fortunate enough to receive "God's blessing" in the form of goods washed up on the beach—that is, cargo from shipwrecked vessels. These goods were naturally always a welcome windfall that bettered the situation of the lucky finders. According to the customary law at the time, the finder was the new owner of any goods found in the water unless the real owner was still alive. These so-called *stranding rights* sometimes meant that finders of stranded goods ensured that none of the shipwrecked individuals survived. In some parts, coastal inhabitants even used fire, lanterns, and other tools to influence the navigation of ships and make them wreck. In both cases, the line to piracy was crossed. However, there were typically no witnesses; and perpetrators seldom faced consequences for their actions—except from the church. Two religious interpretations of such occurrences competed with each other: (1) the finders' view that the shipwreck was God's wrath for the individuals on board and the goods God's blessing for the finders; and (2) the pastors' Christian love and altruism for the shipwrecked individuals. The church was not alone in its demand to change attitudes toward shipwrecked persons as well as the common practice of stranding rights. Although most owners of cargo and ships were not shipwrecked themselves, they held ownership rights even in the case of distress or wreckage. Often, the cargo and the ship itself seemed to be worth salvaging; thus, rules to share the benefits were needed. In 1781, a decree was published in the Imperial Regulations of the Merchant Shipping that defined the proportion of the proceeds from salvaging wrecked ships that would go to the salvager and other stakeholders, depending on the wreck's distance to the shoreline. The same article (#281) of this decree let it be known that lifesaving must be done free of charge out of philanthropy and charity. Those who put themselves at risk to do so would earn special favor and a medal as a mark of distinction (Russwurm, 1865). This second source showed that philanthropy is an honorable value irrespective of what the church says. Still, the church had additional arguments that are more closely related to the living conditions in the poor communities with their family ties: brotherliness. Indeed, the survival of the coastal communities was closely related to fishing and life at sea. The loss of a fishing boat and family members was a real threat for each family but also for the community that needed able bodies to survive. In addition, one thing was the same 200 years ago as it is today: If a fishing boat is in distress, colleagues would sacrifice a good catch to save the lives of other people like them. However, the question remained: Is it acceptable to voluntarily leave the safety of land to rescue shipwrecked individuals and risk one's own fishing boat(s) and men in the rough sea? There is no quick or easy answer to this question.* That coastal inhabitants should be encouraged to treat a sailor's body with dignity after being shipwrecked is well documented in a West Frisian tradition: Sailors or fishermen traditionally wore a golden earring with their initials in the center to make sure their body would be brought home for proper burial if they were found somewhere on the beach.

However, the first brave men who saved persons from shipwrecked vessels— regardless of whether the shipwrecked were from their own community or elsewhere—were volunteer coastal inhabitants. Such individuals would eventually become the initiators of future charity initiatives and local lifeboat associations.

3.1.2 ABOUT MERCHANTS OR MONEYBAGS

A much smaller but wealthier population in the 19th century, just as in the centuries before and after, were the successful merchants in the Hanse towns in Middle and Northern Europe. In the German-speaking regions, they represented a counterweight to established aristocracy and the clergy. In the Hanse towns, it was primarily merchants who influenced communities and shaped policy. Their trading ensured economic prosperity and political independence for most of the Hanseatic cities. Merchants' roles in their respective hometowns were determined by both their prosperity and the offices they held—positions that were extremely limited. Out of their key positions, numerous bills were made by this lobby. Most of the enacted laws concerned trade to preserve the merchants' interests. Their chief interest lay in the security of their commodities and market terms. However, they did not influence all realms of life at the time: Two examples include customs duties that had to be paid for cross-border trade as well as rights concerning shipwrecked cargo. Until the German nation state (German Empire) was founded in 1871 under the Prussian king Wilhelm I, the territory was fragmented into smaller states reigned by dukes, princes, and kings. That constellation interfered with trading, increasing transaction costs and limiting merchants' profits. One of the abstract ideas in the 19th century—to abolish customs duties—was a motivator behind the unification of Germany. The fragmented German territory was also costly in terms of obtaining rights. Specifically, the common practice of stranding rights as well as poor surveillance measures and law enforcement affected the trader's income. Even severe penalties for shipwrecking and piracy had almost no effect on the activities of the coastal inhabitants due to the closely-knit circles of the involved persons and clannishness. However, the smart Hanseatic merchants were aware of the fact that most cargo was not needed in coastal hamlets. Thus, they exercised their influence to make laws for the main emporiums regarding the reclaiming of shipwrecked goods as well as a death penalty for trading stranded or stolen property from wreckage. Still, a uniform legal framework could only be ensured in a united German nation state. Thus, all

---

* This question was a core theme in the traditional literature—for example, those of Nis Randers, a potential rescuer who had to weigh altruism, brotherliness, and charity against responsibility for his own family. At the end of the day, he ended up accidently saving the life of his sorely missed brother.

attempts and initiatives to unify the fragmented German states received strong support from the Hanseatic merchants.

The social structure in the Hanseatic towns was not as close-knit as in other towns
in the 19th century. Social climbing to leadership positions in the town's ruling class
was always an option for entrepreneurs, as well. Economic success and social engagement meant that fast climbers could become part of the functional elites. Likewise,
the established patriarchal elites had to emphasize their social commitment in order
to demonstrate their responsibility toward the community. Holding office for welfare
or care of the sick was highly esteemed when individuals attempted to achieve positions in business and politics (Elmshäuser, 2007).

Both factors made Hanseatic merchants an attractive and obvious counterpart for
founding a charity to promote Germany's unification and ensuring their social standing in their hometowns. Creating a charity called "German Association…" was a
strong political statement, especially because it was not certain that Germany would
actually be unified. Adding the charity's main purpose to the name—that is, to prevent the loss of life at sea and help shipwrecked vessels—offered the merchants a
direct link and fast access to incidents that would otherwise be nearly beyond their
area of influence. This was an incentive that "qualified" the involved persons for both
working pro bono and being potential donors for this charity. However, an important
new concept must be developed and presented to the right people at the right time.

3.2 THE BEGINNINGS OF AN ORGANIZED
MARITIME RESCUE SERVICE*

Slowly but surely, piece by piece, the ideas of humanity and charity became common sense in the 19th century. Several charitable organizations have their roots in
that time. One of the earliest documented initiatives to build up capacities to provide rescue and salvage took place in the area of Memel—today's main seaport of
Lithuania, Klaipeda. In 1802, local merchants stationed a rescue skiff in Memel that
could be manned by sailors if needed. However, this was merely a local campaign
group with additional intentions. One could view the rescue skiff as a solely altruistic
activity. Yet, one of the aims at the time was financially oriented: to invest in salvage
capacities, minimize the loss of one's own cargo, and receive a share of salvage
rewards. Still, we do not wish to underestimate the symbolic effect to the public of
such an action, which became increasingly aware of the destinies of survivors and
the rescuers who risked their lives for them. Although in other European countries,
such as Great Britain and the Netherlands, precursors of today's rescue organizations
were founded in 1824, the situation in the German-speaking area of the Baltic and
North Seas was still nearly the same as that 100 years prior. According to estimates
solely for the German islands in the North Sea, there were approximately 50 ships
in distress per year. Then, like today, large disasters were often a trigger for action.
Two major accidents are mentioned as being decisive for the establishment of an
organized maritime rescue service. First, in a heavy fall storm in 1854, the migrant

---

* We thank Christian Stipeldey M.A., Head of Public Relations at DGzRS, for his content support and
the proofreading of the historical facts.

ship *Johanne* stranded near the Island of Spiekeroog; 84 persons drowned. A second, very serious maritime casualty was the sinking of the French brig *Alliance* after it ran aground off the Borkum reef six years later; the entire crew lost their lives.

In contrast to other accidents, the two described above happened in the presence of independent eyewitnesses. They reported seeing poor rescue coordination, inadequate equipment, and a lackadaisical attitude on the part of the local inhabitants. Certainly, the practice of stranding rights provided little incentive for the inhabitants to assist the rescue. In addition, lacking resources due to poverty made the locals reticent; should they risk losing the only boat they have to earn a living in an attempted rescue?

However, social norms and values had already changed in the German general public. In addition, the public consternation with the shipwrecked increased due to the eyewitness reports. Inspired by those reports, on October 3rd, 1860, the navigational lecturer Adolph Bermpohl named and shamed the sorrow of the shipwrecked and safety situation in German waters in a newspaper article. He also called for founding a private national rescue association like those in the UK and the Netherlands, adding that donations from the German public should be sufficient to finance those efforts. Together with the solicitor Carl Kuhlmay, he distributed a "Proclamation to Contributions for the Establishment of Rescue Stations on the German Islands in the North Sea" on November 20th, 1860, to the desks of Northwest German newspapers. Their appeal inspired Dr. Arwed Emminghaus—editor at the Bremer commercial paper—and Georg Breusing, chief customs inspector of Emden. It was Georg Breusing who initiated the founding of the local "Association for the rescue of shipwrecked at the East Frisian coast" on March 2nd, 1861, as the first charity of this type in Germany. He established rescue stations on several East Frisian islands. In the following years, similar local associations were founded in Hamburg with a rescue station in Cuxhaven and Stralsund as well as Bremen with rescue stations in Bremerhaven and on the island Wangerooge. Additional foundations followed in Lübeck, Rostock, Stettin, and Danzig. Emminghaus pursued the plan to prevent the fragmentation of all the new local associations. He invested three years in negotiations to attain his goal through diplomacy. On May 29th, 1865, most local rescue associations assembled in Kiel to found the German Maritime SAR Service: the DGzRS. Emminghaus proposed Consul Hermann Henrich Meier as the first chairman and dedicated considerable effort to convince Meier to take on the honorary appointment. The Consul from Bremen, one of the founders of the North German Lloyd and chairman of the sea commission of the Bremer chamber of commerce, eventually agreed. Bremen subsequently became the designated headquarters of the DGzRS. Until today, the chairmen of DGzRS were historically Bremen merchants working on a volunteer basis. Thanks to Meier's effort, the patron of the rescue service was and still is the acting head of state—today, the acting Federal President. In addition, all DGzRS operations were and still are exclusively financed by donations without any governmental influence. In the decades following its founding, the DGzRS opened additional rescue stations, which were mainly equipped with gear for beachside rescue, such as skiffs, rocket rope guns, and breeches buoys. Additional advertising campaigns began to attract brave volunteers to begin the altruistic work. After negative experiences with skiffs of the English Peake class as well as the American lifeboat designed by Joseph Francis, the organization eventually developed the standard German rescue boat. Such boats were stationed in a shelter near the shore and then towed to the beach with horses when needed. Each boat was manned with six to eight scullers and one coxswain—to this day, the rescue cruiser's master is called coxswain by tradition. Their volunteer work was as burdensome as it was dangerous. Still, after 10 years of operation, the DGzRS had achieved some successes: 91 rescue stations were operated at the North and Baltic Seas; and approximately 870 shipwrecked persons were rescued during that period. Around that time—in 1875—the lifeboat-shaped donation collection box was mentioned for the first time. Approximately 1,000 boxes were distributed across the coast and inlands. Today, approximately 14,000 are in use across Germany, making the collection box one of the best known in Germany. Twenty-five years after the consolidation in Kiel, the DGzRS operates 111 rescue stations between the island of Borkum in the West and Memel in the East with more than 1,000 volunteer rescue men under command. Since its founding, the DGzRS has become more reliable and safer for its volunteers through in-house technical innovations as well as new technologies and processes. In addition, in-house testing and improvements continuously better the rescue equipment. In 1911, the first motor rescue boat, named *Oberinspector Pfeifer*, was brought into service—an important milestone in the history of sea rescue. Just two years later, the DGzRS operated 14 motor rescue boats—eight custom built as well as six refitted sailing rescue boats—at the key rescue stations. Unfortunately, the early technology was vulnerable; and World War I interrupted the engineering progress. After the start of the war, small but reliable diesel engines were developed, which coincided with a switch from open to closed rescue boat hulls. The war's end caused some changes in the organization of the DGzRS, which had to surrender the stations in the Danzig area and on the island of Rømø; the stations in the vicinity of Memel were operated only provisional. According to international regulations in 1927, the aeronautical distress call "Mayday" was introduced in the maritime sector. The year 1930 marked the first voice radio on board a rescue boat. Previously, electricity generation for radiotelephony was a problem. However, by 1943, all motor rescue boats in the DGzRS fleet were equipped with voice radios. Even the 1930s and early 1940s brought considerable technological developments despite the fact that the political situation under Nazi control caused difficulties for a charity that highlighted its independence from state aids and governmental influences. During World War II, the DGzRS ships were marked with a Red Cross—a sign for additional, war-related events—and worked under protection of the Geneva Convention: Shipwrecked and shot-down soldiers from all nations involved in the war were rescued by the DGzRS fleet. For safety reasons, rescue units near endangered infrastructures were stationed in sea positions to stand by for the rescue of shot-down pilots as well as to be protected against broad enemy attacks. These boats were the first to be manned by professional, paid SAR workers. This was the beginning of a mixed fleet deploying both volunteer-manned and professionally manned boats. A visible constructional change on the rescue boats during this time was the addition of a tower as a part of the superstructure to better the lookout conditions. After World War II, the Allies supported the reconstruction of the DGzRS and permitted—and supported—the organization to continue their operations and fundraising. Still,

some losses were suffered: As a result of the German partition after the war, all rescue stations in the East German sector and farther east were lost—in total, 70 rescue
stations between Lübeck and Memel. Only the 30 stations in the North Sea (German
Bight) and 10 on the Baltic Sea's southwestern coast remained. Because the DGzRS
headquarters in Bremen were destroyed by allied bombardment in 1944, the mission
control was temporarily moved to Cuxhaven. At that time, 14 emergency radio stations were operated by the DGzRS as relay stations for the communication between
the mission control, the coastal radio stations, and the rescue units. In 1952, the
organization's new headquarters in Bremen was opened, which included a repair
shipyard building, a small pier, offices, and a radio room. Three radio operators
provided 24/7 accessability—the roots of which eventually developed into today's
Maritime Rescue Coordination Centre (MRCC). The rescue units, as well, had to
adapt to changing conditions. The aging motor rescue boats were too slow for the
modern maritime traffic, resulting in an urgent need to renew the fleet. The new
generation of rescue boats was customized for various kinds of missions and changing conditions of the region. The speed, for instance, was doubled, ensuring high
speed even in heavy seas. The new boats were also easy to handle in both deep ocean
waters and shallow coastal waters—tailored for working near the mudflats and sandbanks in the North and Baltic Sea region. In the end, a fast, self-righting rescue
cruiser with a piggyback daughter boat was created.

Continuous improvements as well as technical innovations improved the rescue
cruiser concept from the mid-1950s until today. The volunteer rescue boats were also
motorized and always utilized state-of-the-art equipment. After the fall of the Berlin
Wall, the former state-operated rescue stations east of Lübeck to Ueckermünde near
the Polish border returned to the DGzRS. In 2015—the 150th anniversary of the
DGzRS—54 stations with 20 rescue cruisers and approximately 40 rescue boats were
permanently on call. Although the rescue cruisers are manned at all times by three to
seven full-time, professional crew members, depending on the cruiser's size, volunteer
station boats are on standby and can be manned with volunteers following an emergency alert within a few minutes. The current deployment plans to integrate or add
accommodations to some station buildings, if possible, so that the personnel no longer
hast to live in the cruiser cabins. This is primarily an economic consideration but represents a desirable solution from the human factor point of view, as well. Stations with
crew accommodations on shore can operate more efficient units that do not need to
carry four or more bedrooms and everyday necessities, such as cooking equipment. At
locations where this concept is not practical, larger units—from 28 to 46 meters total
length—provide ideal capacity for both SAR demands and life on board.

Similar to other ships, the crew is structured hierarchically. Commanded by a
first coxswain, the positions on board are bipartite. Navigators and engineers share
the main tasks to operate the cruiser. Occasionally, they are complemented by
rescue men, who typically possess some skills in both areas and drive the daughter boat. All crew members work in 14-days shifts: 14 days on board followed by
14 days off work. During their onboard time, the crew is together 24/7. Most time
is spent with maintenance tasks and ensuring that all technical equipment is ready
for use. During these times, the generally strong hierarchy on board relaxes, and
informal organization becomes prevalent. Until now, there has been no reason

to question this practice in terms of if and when it affects mission performance. However, after visiting all professional rescue stations—which we did while preparing our quantitative inquiry—it became clear that each station has unique customs and traditions. The stations' different working and living atmospheres coincide with different assignments dependent on their working region, such as searching for missing surfers near famous windsurfing and kiting spots or salvaging pleasure crafts in a popular maritime tourist destination with sandbanks and shoals. Other stations have a large number of medical evacuations (medevacs) from ferries and other commercial shipping vessels due to their close proximity to main shipping routes. Still others frequently offer ambulance services in structurally weak areas, which is not SAR by definition but is necessary under certain circumstances. Finally, some stations are highly experienced in operations involving fishing vessels because they are stationed in the vicinity of fishing grounds. As a result, different stations possess different areas of expertise according to the unique types of missions that they typically encounter. Thus, applied basic research in this field of practice must take such differences into consideration.

3.2.1 Legal Basis for Suretyship

A set of rules and regulations forms the legal basis for SAR work in general as well as the specific national structure. Although international regulations, for example, by the International Maritime Organization (IMO, 2000a, 2011, 2014a,b) or the International Labour Organization (1991), lay down broad guidelines for SAR activities, national regulations—for example, negotiated with the Federal Ministry of Transportation—determine the structural embeddedness and individual responsibilities of the concerned parties. The international basis for worldwide SAR is the

---

OBJECTIVES AND DUTIES OF THE GERMAN MARITIME SAR SERVICE (DGZRS, 2012)

- Saving human lives in danger at sea and providing first aid
  - Coordinating all actions in emergencies at sea and when assisting missions within the German SAR territory
  - Monitoring VHF channels 16 and 70 for emergency and safety purposes as well as handling distress, emergency, and safety radio calls on VHF in the German SAR region
  - Carrying out preemptive missions to secure potentially endangered vessels and crews
  - Aiding in evacuating crews from seagoing vessels and aircrafts in immediate danger
  - Transporting and providing first aid to the sick and injured
  - Taking any measures to prevent distress and accidents
  - Assisting German vessels and crews in emergencies abroad
  - Assisting units engaged in firefighting if feasible
  - Assisting the Central Command for Maritime Emergencies in disaster management

---

so-called IMO SAR convention (IMO, 2006). The three volumes of the *International Aeronautical and Maritime Search and Rescue Manual*—Organization and Management (IMO & International Civil Aviation Organization, 2010b), Mission Coordination (IMO & International Civil Aviation Organization, 2013), and Mobile Facilities (IMO & International Civil Aviation Organization, 2010a)—include operational guidelines for all levels involved in SAR activities, beginning abstractly with the configuration of an SAR region of responsibility for administrative decision-makers in Volume 1 to search mission coordination guidelines for MRCC staff in Volume 2 to instructions for seafarers who could potentially be involved in a distress situation—both directly and indirectly—in Volume 3. The Maritime Safety Committee of the IMO recommends in their resolution No. 894(21) from November 25th, 1999 that all governments "ensure that all ships entitled to fly the flag of their countries carry on board a copy of Volume 3 of the IAMSAR Manual" (IMO, 2000b, p. 2). The IAMSAR manual is the SAR "bible" whose content is continually reviewed and periodically updated. These international regulations and guidelines are supplemented by national guidelines in due consideration of the national SAR capacities and their integration into the general emergency response organization of the coastal nation. In Denmark, for example, the international regulations were supplemented by the SAR Denmark guidelines published by the Danish Ministry of Defence that is in charge of SAR in Denmark (Danish Ministry of Defence, 2011, 2015). In terms of the German Maritime SAR Service, it is apparent that international regulations, that is, the SAR Convention, primarily addresses coastal national entities and not nongovernmental organizations. In addition, many of the conventions were created in the 1970s—more than 100 years after the rescue organizations in the UK and Ireland, the Netherlands, and Germany began their operations. The main question is how to adjust these long-standing systems to the new international regulations. In the following, we briefly discuss how this was achieved in Germany. It is important to note, however, that such a process would likely look quite different in other nations.

As mentioned above, preexisting communications as well as command and control structures have been operated by the DGzRS since the early 1950s. When the IMO SAR Convention (IMO, 2005) entered into force, the rescue coordination in the German exclusive economic zone was already guaranteed by the organization-specific rescue control room. According to international (Art. 98/2, United Nations, 1982) and German national laws, it is the government's responsibility to make provision for maritime SAR. In Germany, this responsibility ultimately falls on the German Federal Ministry of Transportation. Based on the preexisting structures and resources, the Federal Ministry of Transportation transferred its official duty to provide maritime rescue coordination according to the IMO SAR Convention (Bundesministerium für Verkehr, 1982a) to the nongovernmental rescue organization and designated the DGzRS rescue units exclusively as primary SAR units (Bundesministerium für Verkehr, 1982b). An evident advantage of this solution is that, in addition to the fact that a substantial governmental investment was saved, rescue coordination and operation stayed in one hand. In doing so, the DGzRS (at the time called the MRCC Bremen) had direct access to the rescue units along the coastline and functioned as the point of contact for both vessels in distress and other national rescue coordination centers abroad. In other countries, the MRCC—or Joint Rescue Coordination Centre (JRCC) when it jointly operates with the national Aeronautical Rescue Coordination Centre (ARCC)—can be part of the national armed forces/navy, such as in Denmark, or managed by the responsible maritime authority, such as in Sweden. Meanwhile, the responsible ARCC in Germany is under military command. Although the DGzRS operates only surface units, an agreement concerning mutual assistance with the Ministry of Defense ensures that naval helicopters can be requested for civilian SAR just as rescue cruisers are made available for civilian aeronautical or military SAR operations (Bundesministerium der Verteidigung & Bundesministerium für Verkehr, Bau- und Wohnungswesen, 2001). Furthermore, the MRCC Bremen contains the coastal radio station Bremen Rescue Radio, which monitors the VHF channels 16 and 70 to respond to distress and emergency calls inside the German SAR region of responsibility. In 2003, the Federal Ministry of Transportation and all five German coastal states funded a unified command system for major maritime incidents in German waters—the Central Command for Maritime Emergencies (CCME)—to coordinate all measures concerning pollution control, firefighting, medical assistance, and emergency towing. The national maritime SAR manual *DV 100 See* (Deutsche Gesellschaft zur Rettung Schiffbrüchiger & Havariekommando, 2004), jointly developed in 2004 by the DGzRS and CCME, ensures that all SAR forces and law enforcement authorities at sea and on land utilize compatible procedures in their work. Where and how the MRCC Bremen and primary SAR units (rescue cruisers and volunteer boats) of the DGzRS collaborate with the unified command system in the case of a major maritime incident is regulated by a 2002 amendment to the 1982 agreement between the Federal Ministry of Transportation and the DGzRS (Bundesministerium für Verkehr, Bau- und Wohnungswesen & Deutsche Gesellschaft zur Rettung Schiffbrüchiger, 2002).

3.3   WHY IS MARITIME SAR A SUITABLE FIELD FOR DECISION-MAKING RESEARCH?

Decision-making research is primarily conducted in laboratory settings due to the ability to control for confounding effects, and thus obtain highly valid and reliable data, that such a setting offers. However, such methods are associated with various weaknesses concerning practicality and the generalizability of results to real life (Lipshitz, Klein, Orasanu, & Salas, 2001). In addition, because most decision-making research is conducted at institutes of higher education, the participant pool in these laboratory settings typically consists of students. Student samples are presumably adequate when a study's objective is related to or near students' real-life experience. However, this is presumably not the case when the studied effects focus on real decision-makers who are experienced professionals working under very different environmental conditions, such as in harsh natural environments using a self-contained breathing apparatus in the face of great danger. It is nearly impossible to transfer such decision-making conditions as well as an adequate number of experienced professionals to a laboratory setting to then reliably examine how such decision-makers would actually decide in practice. Several human factor studies in the aviation domain, for example, have shown that simulators can be used as a laboratory setting for applied research. This and the high relevance of safety in aviation—errors or mistakes may result in hundreds of casualties—makes this sector one of the most extensively examined in applied research. Such studies, as well as those situated in other high-reliability organizational contexts, for example, firefighters and paramedics, are familiar to executives and managers in commercial business who look for inspiration, metaphors, and examples to handle critical situations in their daily work or in preparation for a potential major crisis. Practical tools have been transferred from the cockpit to the meeting room and from a paramedic's workplace to human resource managers (Weick & Sutcliffe, 2007; Weick, Sutcliffe, & Obstfeld, 1999). High-reliability organizational studies are often so successful because they are applicable to both the practical fields in which they were conducted and other domains in which best practice examples are much appreciated. What makes maritime SAR a suitable context to study decision-making questions? What insights can we expect that could not be obtained in student sample–based laboratory research?

High-reliability organizations are built around the idea of reducing errors to an absolute minimum under constant awareness of near misses, while also being prepared for critical situations and possible accidents. High-reliability organizations are diverse; and each sector within the organization is as unique as the organization itself. However, first and foremost, such organizations share the idea that changes should only be made if one can objectively show that they improve the reliability of the organization's operations. Oftentimes, the environmental and organizational context of high-reliability organizations is thus quite stable—technologies change relatively slowly; and the organizational environment also tends to relatively consistent. However, the situations in which these organizations operate are quite dynamic and oftentimes characterized by an immediate danger for life, health, property, and/or the environment. As a result, they are of immediate importance to both the immediately involved stakeholders and society as a whole.

These conditions cannot be fully reproduced in a laboratory setting, especially one that does not involve nonprofessional decision-makers. As discussed in Chapter 2, the decision modes we examine in this book are deeply rooted in knowledge and routines, which are both field-specific.

Another important reason to look at decision-making in high-reliability organizations in general, and maritime SAR in particular, lies in the specific character of decision-making in these fields. First, decisions tend to be highly visible—that is, an observer is typically able to clearly specify when a decision has been made and by which person. This is not the case in more convoluted or socially burdened decision processes, for example, in boardrooms. Thus, we make use of this characteristic in the context of video-based shadowing studies, which we will discuss in Chapter 5. Second, the environment tends to provide immediate feedback to specific decisions, in that decision-makers can almost immediately observe the outcome to which the decision led. Again, this is not true, for example, in the case of management decisions with which the outcome may not be visible until months or years later as well as with which it is oftentimes not clear whether the outcome occurred because or despite of a particular decision. This is an important aspect that helps decision-makers identify their own decisions, which, in turn, is critical for the survey methodology we applied to study decision modes. We will report these findings in Chapter 4. In sum, we believe that high-reliability organizations, in general, and maritime SAR, in particular, are excellent settings to study our research questions. We believe that we have chosen the best available methods to answer the questions "How do professional decision-makers decide in maritime SAR?" and "What impact does the complex task environment with which they are faced have on their decision-making?"

In contrast to emergency assignments on land, rescue at sea implies that the rescue crew must sail its own ship into the same conditions in which the emergency situation emerged. In some cases, this is not an issue, as the environmental conditions are not always the trigger of a maritime incident. However, more often than not, the environmental conditions are at least partially the cause. Most rescue missions encountered by the DGzRS are not SAR, per se; instead, if a rescue unit acts quickly and assists immediately, the assignment primarily involves providing technical assistance. Still, rescue work always means working in a dynamic environment in which conditions can change quickly. If the rescue cruiser cannot assist quickly enough, which almost certainly occurs if the case is far away, changing weather conditions may make the mission unpredictable. For example, with increasing wind and a subsequently growing sea state, a minorly damaged and disabled vessel adrift at sea may quickly become a serious casualty. All maritime missions are susceptible to the environmental dynamics at sea. The longer the mission, the higher the potential influence of the environmental dynamics to the casualty and the mission itself are. Minimizing the time that is necessary to locate and reach a person or vessel in distress is the first step to prevent the loss of life at sea. In expansive SAR regions of responsibility, the first strategic decision is what type(s) of rescue units (surface and/or air units) are appropriate to deal with the demands of a particular case. Is there a chance of providing a safe return to port on board the distressed vessel in order to protect life, the environment, and property, or can only the lives of the shipwrecked persons be saved? This and other operational and strategic decisions are made by the search missions coordinators—professional decision-makers—working at the MRCC on land. These individuals are mission leaders working in a unified command system that is quite similar to those working in disaster management in other contexts. Extensive research has already been conducted on this topic. Thus, how command systems work effectively in diverse contexts is well understood. The more interesting aspect is thus the work done on scene—that is, in the danger zone, where workers are physically and mentally faced with changing conditions and unpredictable situations. It is very challenging to build an effective SAR organization when resources are constantly scarce. Personnel from different organizational backgrounds, working experiences, and organizational cultures must cooperate to save lives. Hundreds of decisions are made in maritime disaster response organizations—from small decisions, such as when a deckhand decides to carry a penknife in his pocket, to big decisions, such as when an on-scene coordinator recommends that a vessel in distress be evacuated—and each contributes to the successful handling of a disaster. Such a decision environment is markedly different from those in aviation, in which two or three individuals in a cockpit of a passenger airplane follow standard procedures under extreme time pressure. In SAR decision-making, there is also time pressure. However, SAR workers are typically not dealing with their own emergencies but rather those of other parties. Thus, the work is about placing one's own safety on the line to help others and acting as a small part in a large picture. Such situations can only be handled effectively when all involved parties successfully fulfill their roles instead of pursuing their own objectives.

In terms of commercial business, similar decision situations can be found in post-merger situations, public–private partnerships, and restructuring enterprises. Our aim in this book is to provide an approach, examples, and ideas regarding how one should deal with the complexity of uncertain situations in maritime SAR as well as encourage the generalization of our findings to other contexts. We focus on key persons fulfilling exceptional missions and analyze the decision situations with which they are faced. Situations that require creativity to find practical solutions, prioritizing the next steps, measuring the competencies and capabilities of the cooperating parties, and, at the end of the day, making suggestions to manage the necessary ad hoc cooperation in a complex task environment—this is what makes our considerations unique.

In the following, we will describe in detail several defining characteristics of work in maritime SAR.

3.3.1 RELEVANCE OF PROFESSIONALISM

As mentioned previously, the severity of a maritime incident varies from case to case. Sometimes it may be a leisure craft, for example, a sailing yacht, that is sending distress calls via radio from the open sea because of engine problems. If the yacht is stable and no additional problems subsequently occur, this is not an SAR case. Depending on the skipper's sailing experience, he or she can sail the yacht to the next harbor and then ask for towing assistance when they reach the nearest port. Still, although potentially harmless, such alerts must be taken seriously: Who would call for immediate help without a reason? SAR workers must assume that someone is overwhelmed with the situation at hand and requires assistance to fix the problem. A failure to treat situations such as these seriously could potentially cause them to turn into actual SAR cases. Depending on the persons involved, the specific situation might be a mere trifle or the beginning of disaster. Every year, situations occur in which the only person on board a vessel who is able to sail the boat becomes ill, injured, or goes overboard, and none of the remaining persons know what to do next (Investigation report #A-08/2010, *SS Kelbo*, March 29, 2009, cited in Federal Bureau of Maritime Casualty Investigation, 2011). Thus, rescue workers must brace themselves for being faced with the helpless—a situation that requires them to solve all problems using their own resources.

Support by persons on board a pleasure craft in distress is typically not to be expected. The rescue workers' advantage is their professional qualifications (see Figure 3.1). They must be well-skilled seafarer, able to judge the conditions and

FIGURE 3.1 On-scene coordinator working on a small search and rescue unit. (To preserve the privacy of all decision-makers described and depicted in this book, we anonymized all photos.)

circumstances of a distress situation and be just as creative as practical in order to identify appropriate solutions. The more rescue workers have seen and experienced themselves—for example, sea state, storms, rolling and pitching vessels, critical situations on board, etc.—the better they are able to estimate the risks associated with their ideas and tactics. In addition, a wealth of experience allows rescue workers to empathize with the (sometimes counterproductive) activities and behavior of persons in distress. Concerning large-scale accidents, the coxswain's experiences should be more comprehensive than those of a recreational skipper.

Knowledge of stability calculations as well as basic knowledge of dangerous cargo and hazardous materials, maritime law, and crisis management will help rescue workers accurately assess the situation. In addition, it is necessary to have a crew member onboard who is an experienced (mission) leader—a skill that master seafarer in leading positions typically internalize (see Figure 3.1). In light of the fact that medical evacuations are routine but not daily business in the SAR domain, qualifications and practical experiences as a paramedic are also of key importance. Some injuries that patients at sea may have suffered could potentially overwhelm a first-aider. To be prepared for such situations, rescue workers are required to complete internships in ambulances and emergency rooms for several weeks as part of their training. The amount and variety of reference experiences serves as a basis for personal preparedness for SAR tasks.

This is why our research focus is on professional maritime SAR personnel. Based on their knowledge, experience, and skills, more mission responsibilities are handed over to them in a large-scale maritime disaster than to volunteers. Focusing on professional workers bridges the gap to commercial organizations and enterprises; it is difficult to compare professionals in an enterprise at the same level with humanity- and passion-driven volunteers in a charity organization. The difference between the two are great. However, the link between professionals, regardless of their organization, also appears obvious.

3.3.2 WORKING WITH AND AGAINST UNCERTAINTY

There are diverse means with which rescue workers can become informed about a vessel's distress situation: Nearly all types of telecommunication—radios on different frequencies, telephones, etc.—are used by different persons who are more or less deeply involved in the situation. Redundant automated alert systems—for example, emergency position indicating radio beacon (EPIRB)—provide additional alarm functions in case of quick sinking. A limited amount of digital information is transmitted by these devices, which differs depending on the technology used: an identification number (maritime mobile service identity, MMSI), the last known position, and, if the alert was made via digital selective call (DSC), information regarding the type of distress, for example, fire on board or abandoned ship. However, distress information is always limited, and the facts—in the form of verified information—are even more limited. Depending on the information source, rescue workers' impressions of the situation as a whole can be rather vague. When the MRCC receives information on a perceived maritime incident, the SAR mission coordinator must investigate the case immediately.

The first decision is whether all available information indicates a ship or persons in distress or not. The IAMSAR manual Vol. II distinguishes between three initial action stages: the uncertainty phase, alert phase, and distress phase. These phases must be declared by a rescue coordination center for each specific situation. The more information and facts the search and rescue mission co-ordinator (SMC) has, the more quickly a phase can be declared and the required actions taken. However, reevaluations of the action stage are often required when new information surfaces or unspecified information is verified. SAR units are often used to gather information on scene or verify information provided by third parties, such as shipping agencies or relatives on land. Thus, the SMC who is in command of the supposed rescue mission must investigate all directions and utilize ships in the vicinity of a casualty to check the facts and initiate immediate measures to push the rescue forward. This typically means initiating a search for the distressed vessel. Most information obtained must be checked by the SMC or a SAR unit on behalf of the MRCC. The constant collection of information and alignment between the decision-makers on scene and the decision-maker in charge in the MRCC makes the SAR domain very interesting for research, as the framework for decision-making may change at any time from one decision to the next. From our perspective, the behavior of the decision-maker on scene has the greatest relevance for research: These persons are faced with a maximum of environmental factors that may influence their decision-making. We are interested in findings concerning the use of varying decision modes under different degrees of (un-)certainty or limited facts.

3.3.3 LATENCY AND COMMUNICATION

Another important circumstance in the professional maritime SAR domain that we would like to point out is latency. Especially during the winter months, the mission frequency is very low, with less than one in 14 days during the slowest periods. It can be very frustrating for the crews to be on standby all the time. If the SAR duties are in the hand of a coast guard, there are other official duties to be fulfilled as long as no emergency situation occurs. However, for professional, full-time rescue workers, living together in a cramped rescue cruiser for two weeks at a stretch can be difficult. These circumstances are highly interesting from a social–psychological perspective. Exercises, patrolling their area, and collectively performing maintenance tasks are good means to prevent boredom and subsequent problems between crew members. However, SAR cases that occur in the wintertime are often more difficult to handle, safety-critical, and tragic due to the poor environmental conditions in comparison to the warmer months. This makes SAR work more demanding, as rescue workers must instantaneously switch from standby to mission mode. Experienced seafarer, for example, reported experiencing extreme seasickness after being stranded in the harbor for a few days due to embacle. However, latency does not only mean that no missions take place. Latency can also refer to procedures that are not regularly practiced and technologies that are not regularly used. An example is radio communication. Typically, most radio communication near the German coastline is conducted in German. Even most ships and pleasure crafts from the bordering countries, such as the Netherlands and Denmark, speak German on the VHF radio. This convenient circumstance means that English communication on the radio is generally not well practiced. Although this is not an issue most of the time, it can be highly problematic if a party becomes involved in SAR activities that does not speak German. In contrast to other domains, the maritime domain has its own safety language—Standard Marine Communication Phrases (SMCP)—a deliberately reduced language based on a set of vocabulary and phrases that cover everything concerning shipping (IMO, 2005). These phrases must be used as often as possible in order to facilitate effective communication in multilateral rescue missions, which is of key importance to a mission's success. Because it is not possible to obtain routine experience in English radio communication in daily work, rescue workers must train this skill as often as possible.

3.4 RESEARCH APPROACHES

It is the nature of SAR missions to occur unpredictably following accidents. It is therefore nearly impossible to predict when a mission will occur. Empirical studies in this field thus face several challenges: Where, when, and with whom should data be collected? How can we make sure that we observe the relevant environmental factors that may affect the decision-makers?

For our studies, we were interested in decision-making in challenging task environments (see Chapter 1). However, professional work oftentimes consists largely of routine occurrences, and maritime SAR is a prime example of this. Complex assignments occur infrequently. Even when they do occur, it is not clear where and when. Thus, it is impossible to predict beforehand which vessels will see action. As a result, we decided against periodical observations on random rescue cruisers, as the chance of observing actual rescue missions using such a method is presumably rather low. Instead, we utilized different approaches to ensure that we observed as much actual SAR activity as possible. Specifically, we combined four quantitative, survey-based studies—the results of which are discussed in Chapter 4—with three qualitative studies, which form the focus of Chapter 5. Overall, our studies capture decision-making in different grades of realism. One quantitative and one qualitative study were conducted in a ship simulator. In addition, we observed SAR professionals with different degrees of experience during two live SAR exercises, which resulted in one qualitative and two quantitative studies. The third qualitative study involved participant observation in real-life SAR missions in the Aegean Sea in the context of the IMRF "Members Assisting Members" joint mission east of the island of Lesvos. Finally, we applied survey methodology to capture decision-making in real-life, everyday missions. Together, the approaches represent an international and interorganizational framework. All our studies are at least bilateral (simulator study) or even multilateral under participation of more than two nationalities/rescue organizations. In the following sections, we outline the four different approaches.

3.4.1 SIMULATION STUDIES

Simulators are well established in the maritime domain. For decades, they have been used to train bridge personnel handling crucial maneuvers in the harbor or collision avoidance at sea. The simulation is basically a mathematical model that predicts the ship's behavior under defined conditions. In a simulator cubicle, which is designed to look like a ship's bridge and equipped with the necessary navigational gear, one finds the human–machine interface between the mathematical model and the trainee using the simulator. Instructors can control, direct, and define all relevant environmental conditions as well as other ships existing in the training scenario. These so-called ship-handling simulators are operated by navigational schools and colleges, shipping companies, and commercial training providers (see Figure 3.2).

We used a modified ship-handling simulator to conduct our experimental approach. Emergency response is always a communicative action between several parties. Thus, for SAR purposes, a simulation is required in which the units can interact in one and the same scenario. For this purpose, the DGzRS operates a networked Global Maritime Distress and Safety System* simulation in a SAR simulator that consists of five interlinked ship-handling cubicles, each with an integrated communication device. The cubicles can be observed and controlled from an instructor center located in a separate room. An obvious advantage of using simulations for training and research is the ability to control all influencing factors, ease of observation during the exercise, and the repeatability of each scenario. We embedded our experimental approach into an international SAR simulator exercise and, on a separate occasion, an on-scene coordinator refresher training. This

---

* IMO, 2015.

FIGURE 3.2 A cubicle in an interlinked ship-handling simulator.

ensures that exclusively experienced SAR professionals took part in the study. Each cubicle was manned with a maximum of three persons with the exception of one cubicle that could seat up to four, resulting in a maximum of 16 persons per study. Each scenario lasted approximately one and a half to two hours per session. The environmental factors remained constant during the scenarios. In our study design, participants answered questionnaires in two steps. First, they answered questions assessing the general decision style preference of each participant. While the simulation was running, the groups in the cubicles were recorded by video camera. After the scenario, a postmission decision-making questionnaire was handed out to the participants to gather data on the main decisions each crew made, including, example, how decision-making was conducted and what factors may have influenced the decision style used. For the evaluation, the premission and postmission questionnaires were triangulated with the data generated by the parallel observations. We used this method of separating decision-maker and decision-making questionnaires for all survey-based studies.

The simulator environment has strong advantages in terms of controlling circumstances and assignments, yet it obviously lacks the real-life character we defined as important in the discussion above. Therefore, our next study examined large-scale, open-sea exercises.

3.4.2 SHADOWING AT BALTIC SAREX

Rescue missions for large-scale, maritime disasters are rare but not improbable events. The high traffic density combined with the comparatively small SAR regions of responsibility adjacent to one another in the Baltic Sea increases the potential that serious accidents may rapidly become a multinational case. As a result, the Royal Danish Navy gradually developed an initially military SAR exercise for one of the world's largest annual civilian SAR exercise weeks: the Baltic SAREX. Participants—naval and civil rescue organizations as well as maritime and law-enforcement authorities—from the Baltic rim and France send surface and air units to the Danish island of Bornholm, around which the Baltic SAREX takes place, typically in early May (see Figure 3.3).

The island has extraordinary conditions to run a live exercise on that scale. It is located quite centrally in the Baltic Sea within easy reach for all participants. It has an airport and harbor facilities that are large enough to host more than 20 rescue units of all sizes. The Danish military owns several facilities on Bornholm that are necessary to organize the logistics of such an event, including vans for crew transfer, barracks for accommodation, exercise equipment, such as obstacles for rescue divers, and the local Maritime Surveillance Center South—a subcenter to the Danish Joint Rescue Command Center in Århus—which functions as the rescue center during the exercise. Furthermore, the hosting Royal Danish Navy was always able to

FIGURE 3.3 Vessels off the coast of Bornholm during Baltic SAREX 2013.

arrange more than 100 role players for the live scenarios. Together, these conditions result in a highly realistic exercise that is highly valued by all participating nations.

Several additional factors related to the Baltic SAREX are worth mentioning: A local ferry company that operates ferry lines between Bornholm and Germany as well as Bornholm and Sweden typically provides a ferry and crew for one day to run a mass rescue exercise at sea with the ferry acting as the vessel in distress. The mass rescue scenario is tailored for ferries carrying hundreds of persons as well as cars and trucks with different cargo, which is among the most complex scenarios SAR professionals can expect to encounter in the Baltic region. The island's topographical dimensions mean that there is always a lee place nearby, so that tides, current, and wind do not affect the scenarios as much as would be the case at other locations. This provides staff the freedom to design meaningful training conditions. Although it is possible that the environmental conditions are too extreme for some units, the effect is more limited than in other Baltic Sea regions.

Planning an event as complex as the Baltic SAREX is a demanding challenge. In all the years of the Baltic SAREX, the Danish Navy provided well-skilled lieutenants to lead the exercise planning—a time-consuming and also political task, as the diverse expertise, expectations, and requirements from the participating nations must be taken into consideration. In order to more evenly distribute the organizational workload, an International Planning Team (IPT) was formed for the SAREX 2014 and 2015. This team consisted of SAR personnel from three countries—Denmark, Sweden, and Germany—as well as three kinds of SAR organizations: military, maritime authority, and nongovernmental organization. Latest inventions, related technologies, procedures, and SAR tools were well researched by the IPT and transferred into an exercise schedule. A planning conference that took place several months prior to the exercise—hosted by one of the participating nations—was held to determine which units would be sent as well as which contributions the participants could make to the upcoming exercise. Subsequently, the IPT met to create scenarios matching the capabilities of the potentially available units. The main objective when creating a scenario is to always provide "action freedom," which means that a general scenario is given, but participating units are free to make their own tactical decisions without any intervention by the exercise staff.

This results in a maximum of contingency in progress. All relevant information was transferred into a so-called exercise order (EXORD)—a guideline for all exercise days. The EXORD consists of information on the harbor, time schedules, exercise groups, ferry traffic, points of contacts, and much more. The basic structure of the exercise was relatively constant over the past years. The units arrived on Bornholm the Sunday before the exercise starts. The exercise then began on Monday with academic workshops for the commanding officers and seconds in command in a conference facility near the harbor.

At the same time, practical workshops on alternating topics were provided for crew members who did not have to remain on board their vessels. On Tuesday, the "simple SAREX" was conducted in three groups, which was a warm-up exercise for the coxswains and commanders to adjust to on-scene coordinator (OSC) roles. Various small scenarios, following in close succession, required communication and developing SAR organization for a specific emergency case. These scenarios were debriefed in the morning the following day. The scenarios became more extensive on Wednesday, when all units had to participate in two or sometimes three live scenarios of greater difficulty with target vessels in distress and role players requiring medical attention on scene.

All this prepared the attendees for the mass rescue exercise that took place on Thursday. This exercise lasted the entire day. Air and surface units collaborated in this scenario, transferring resources, personnel, and role players playing casualties or injured persons. All activity was observed by umpires, which enabled a proper debriefing afterward, and also ensured actor safety during the exercises. The final debriefing took place on Friday and lasted several hours to reconstruct the overall picture from all perspectives. At the end of the day, a beer call marked the end of the SAR-related activities in the exercise week. Finally, the exercise week ended on Saturday with social activities, such as a soccer competition, golf, and a highly popular bowling match. The entire event is also organized to encourage exchange between the nations as well as demonstrate and advance new technologies. Thus, the exercises also involve the exchange of crew members, for example, an officer or cadet, with another vessel for one day to provide interesting insights into the practices of other organizations.

We had the unique opportunity to obtain full access to this exercise in two consecutive years. We were able to conduct video- and survey-based studies in 2014 and 2015—the final two years of the exercise before it was cancelled by the Danish Navy pending further notice due to the high cost. One of the authors also became a member of the IPT, which provided us with all the necessary information to prepare a survey for this setting. The idea to use the exercise for research was welcomed by the host and participating nations; all were aware of the limited academic research available on SAR as well as the difficulties associated with collecting practice-oriented data.

For our video-based research, we decided to focus on the mass rescue exercise—the highlight of Baltic SAREX. This scenario is extremely realistic with the aid of all available forces. To cover the mass rescue operation from different points of view simultaneously, we placed cameras in key areas and equipped the potential key decision-makers with body cameras—that is, the designated OSC and a second key person, the so-called local incident coordinator* (LICO). A LICO and his staff were requested to support the ship's master and the crew of the distressed vessel by the OSC. The LICO—typically the second in command on the OSC vessel—is more familiar with the local emergency response capabilities and SAR procedures in the territory in which the accident takes place than the crew of the distressed vessel.

Furthermore, the LICO is a person with the OSC's trust, tasked with making a proper assessment of the situation on board as well as providing command and control on board in the case that no one else from the crew can fulfill these duties. The role of the LICO was interesting for several reasons: The LICO and his staff are those who are in the danger zone. They are directly faced with the damage, an unclear and perhaps confusing overall situation, and potential danger to themselves by being on board a possibly insecure vessel. A second interesting dimension to the LICO is that the procedures and ideas behind this structure were quite new at the

---

* Møller, 2014.

time of our observations. Thus, our investigations could contribute to evaluating and refining this new concept for SAR practice. We explore the LICO's role later in Chapter 5, in which the case study is presented. One of the authors also accompanied the LICO as an observer to identify other influencing factors on decision-making and maintain the cameras.

The recordings were edited afterward by a professional video editing service—that is, time synchronized and placed on a single screen—and transferred to an analysis video for further evaluations by SAR experts. These experts dispelled a remaining concern regarding researching a live exercise instead of a real mission: The issue of immersion experienced by the participants. The decision-makers know that the setup is an exercise and not a real case. However, the social pressure on the decision-makers is nearly the same as in a real mission. Real SAR missions typically take place somewhere out of the public eye; the pressure on the decision-makers originates from their responsibility toward the persons in distress. In a live SAR exercise, the pressure on the decision-makers comes from being surrounded by SAR experts, instructors, and international colleagues. Poor and delayed decision-making will influence the scenario in general so that the factual pressure as well as operational time pressure are comparable to real-world occurrences.

In addition to the video data, we also collected survey data during the Baltic SAREX. In 2014, we collected survey data during all open-sea exercises, whereas we focused on the most challenging exercise, the mass evacuation, in our survey study in 2015. We will present the methodological details and the results in Chapter 4. Whereas the video-based studies focused on the OSC and LICO, the surveys examined the decisions of all rescue personnel involved in the exercise.

3.4.3 PARTICIPANT OBSERVATION: FULL PARTICIPATION

The third SAR study we conducted was on a SAR assignment in the Aegean Sea. As a consequence of war and violence in the Middle East, Eastern Africa, and ongoing conflicts in Afghanistan, a massive migration began from these regions to the European Union. In 2015, the migration had a temporary climax with thousands of people passing the maritime border between Turkey and Greece daily. Despite all efforts by the local governments on the island, authorities and volunteers from diverse NGOs—both national and international—as well as the incredible support of the islands' population, the death of people during the short but dangerous passage could not be prevented.

Two Scandinavian SAR organizations decided to support the necessary maritime SAR activities in Greece. The Norwegian Rescue Society (RS) offered one of its rescue cruisers for charter to the Greece Ministry of Interior, which provided this unit to the European Border Protection Agency (Frontex) to have a proper unit for border control in the Aegean Sea. This was done well knowing that migrants' registration at sea is nearly impossible. Still, an appropriate rescue unit was necessary to fulfill the additional obligations according to international law. The rescue cruiser was commanded by a border guard officer sailing with a crew of Norwegian rescue workers and stationed in the port of Mytilene/Lesvos, the island with the highest density of migrant arrivals.

The Swedish rescue organization acted differently: One of the largest publishers in Sweden contacted the Swedish Sea Rescue Society (SSRS) and provided donations from its employees that were doubled by the company to support a SAR mission in Greece. Starting with these donations, a mission on the island of Samos was launched in October 2015. Although not the highest frequented, Samos was the most dangerous location along the eastern Greek border.

As these operations began, the tense situation in the Mediterranean appeared on the agenda of the International Maritime Rescue Federation (IMRF). Subsequently, European trustees of the IMRF started a fact-finding mission to obtain an eyewitness overview of the situations on the affected Greek islands and explore options for fruitful support. Due to the high pressure on their border, limited SAR capacities along the eastern Aegean islands—Lesvos, Chios, Samos, and Kos—and the colder water temperatures during winter and spring, the Hellenic Coast Guard and the NGO Hellenic Rescue Team sent official requests for assistance to the IMRF and the DGzRS as well as other well-established professional SAR services and lifeboat institutions in Western Europe. The rescue organizations decided to initiate a joint support mission headed by the IMRF.

Under the heading "Members Assisting Members," organizations from Middle and Northern Europe supported their colleagues in Greece, providing equipment and training to build up a resilient SAR organization in the Aegean Sea. Temporarily—until the local SAR structures could be established—rescue units and personnel were delivered to the hot spots in the Eastern Aegean Sea to assist the local emergency response. Referring to the request of the Greek authorities, a DGzRS rescue cruiser was sent to operate from the island of Lesvos.

Preparing for this first mission outside the familiar territory of operation both in terms of geography and mission focus and scope, the headquarters of DGzRS aspired to collect as much information as it could to select suitable crew members, provide useable equipment, and create a base on Lesvos that satisfied local requirements. Numerous lessons learned by the Swedish "yellow boats" were shared with the organizations while planning their part of the joint mission. It was obvious that the expected missions in the Aegean Sea would differ from the current knowledge and experiences in SAR service. Instead, a mix of professions—SAR workers and lifeguards—appeared useful in order to cover the maritime SAR operations as well as the rescue of persons from the rocky coast, which is better done by experienced personnel familiar with swimming under difficult current and tide conditions.

The so-called *Lesvos crew* was composed of 12 SAR workers and four lifeguards. Additional Greek colleagues from the HRT supported the crew with on-the-job training. Each team member was stationed in Greece for two weeks. Each week, eight crew members were rotated out, meaning that at least half of the crew at any point in time was experienced in the operations and (new) procedures practiced in Greece and could teach those to the new workers. All the logistics and necessary negotiations with the local stakeholders were managed by the station manager—persons with a broad range of working experience abroad, who were sensitive to the various needs and demands.

Two persons were selected to alternate the station manager position in a four-week cycle. All positions in the Lesvos crew were allocated on a voluntary basis regardless of whether the potential crewmembers were otherwise permanently employed on a rescue cruiser or worked as a volunteer. However, basic requirements regarding physical fitness, vaccinations, good command of the English language, and some medical requirements were monitored by the company medic. Due to insurance concerns, each applicant had to have a German Medical Certificate—a requirement for professionals working at sea. All candidates had to attend special sea survival training.

This training was tailored to make the applicants from both organizations familiar with emergency procedures on board a rescue cruiser and the use of safety equipment as well as to select persons who would potentially cope well with the situation in the Mediterranean. Inspired by the experiences of their Swedish colleagues, a mandatory training on the resuscitation of babies and toddlers as well as a refresher on resuscitation using automated external defibrillators was implemented. Persons who fulfilled all requirements were enlisted in an online database and had to commit their time for the next months. Based on that data, the crew was put together by the headquarters. Operations in the waters around Lesvos were scheduled to begin on March 14th, 2016. Originally, it was planned that one of two standby rescue cruisers—used while another cruiser is in a shipyard for maintenance—would be delivered to Greece for the duration of the joint mission. However, this was not necessary as the owner of a former rescue cruiser that was decommissioned by DGzRS two years earlier planned to help in the same area by establishing his own charity campaign. However, he did not yet have a qualified crew. Thus, he provided the private cruiser to the DGzRS for the duration of the mission. Before its transfer to Greece, the ship was refurbished and equipped with state-of-the-art equipment. An additional container was packed with technical equipment, accessories, and supplies. Finally, a van with mobile facilities was sent to serve as a temporary rescue station in Mytilene.

One of the authors passed the crew assessment and was selected as a member of the Lesvos crew. From this inside perspective, observations were made, diaries written, secondary data collected, and videography conducted to capture as precisely as possible the operations in these exceptional mass rescue operations. Obtaining a deeper understanding of decision-making in practice was one of the aims of this study. Additional aspects were a focus on MRO procedures, continuous improvements, and knowledge management.

3.4.4 Survey Studies during Regular Missions

Finally, we collected data during regular SAR missions in the German SAR zone. As our focus was on complex task environments and we were already aware that most missions did not qualify as involving complex tasks, we began this study by asking experienced personnel to compile a list of all activities they face during their work. Then, we asked other experienced personnel to rate these tasks in terms of their unpredictability, importance, and danger. We then defined those tasks that were consistently ranked highly on all three dimensions as complex tasks.* This procedure resulted in 330 individual tasks, eight of which were subsequently defined as complex.

---

* We borrowed this approach from Colquitt, LePine, Zapata, and Wild (2011).

We then visited all 22 rescue stations with professional crews and asked all crew members to fill out decision-making questionnaires (as in the other survey studies described previously). We then collected those and handed out the decision questionnaires, which the respondents were asked to fill out after each mission that involved at least one of the eight complex tasks. In addition, we read the status reports from all stations on a daily basis to identify missions that might have included tasks meeting our definition of complexity. If we identified one, we called the respective rescue crew and asked them to fill out questionnaires. This study ran for 18 months. In the end, the outcome was somewhat disappointing in terms of quantity, as there were indeed very few missions that qualified as complex tasks. Therefore, we focused our analyses on the other data sources and used the real-life data as complementary evidence.

Taken together, we were able to build on a broad and very rich stock of data, which allowed us to conduct a variety of in-depth studies both quantitatively, with the aim of deriving generalizable insights into how decision modes are used, and qualitatively, focusing on specific, in-depth examinations of decision-making in complex task environments. We will present the survey-based results (quantitative studies) in Chapter 4, whereas the video-based qualitative data are the subject of Chapter 5. In Chapter 6, we will discuss the results in a common context and derive recommendations for practitioners and trainers in and beyond the field of SAR and high-reliability organizations.

REFERENCES

Bundesministerium für Verkehr. 1982a. *Bekanntmachung des Internationalen Übereinkommens von 1979 über den Such- und Rettungsdienst auf See.* Bonn: Bundesanzeiger Verlagsges. mbH.

Bundesministerium für Verkehr. 1982b. *Verwaltungsvereinbarung zwischen dem Bundesministerium für Verkehr und der DGzRS über den Such- und Rettungsdienst.* Bonn: Bundesanzeiger Verlagsges. mbH.

Bundesministerium für Verkehr, Bau- und Wohnungswesen, & Deutsche Gesellschaft zur Rettung Schiffbrüchiger. 2002. *Zusatzvereinbarung zu der Vereinbarung zwischen dem Bundesminister für Verkehr und der Deutschen Gesellschaft zur Rettung Schiffbrüchiger über die Durchführung des Such- und Rettungsdienstes in Seenotfällen vom 11. März 1982 (Verkehrsblatt 1982, Seite 191, Nr. 99) über die Kooperation zwischen der Deutschen Gesellschaft zur Rettung Schiffbrüchiger und dem Havariekommando.*

Bundesministerium der Verteidigung, & Bundesministerium für Verkehr, Bau- und Wohnungswesen. 2001. *Verwaltungsvereinbarung über die Zusammenarbeit auf dem Gebiet des Such- und Rettungsdienstes für Luftfahrzeuge und des maritimen Such- und Rettungsdienstes.*

Colquitt, J. A., LePine, J. A., Zapata, C. P., & Wild, R. E. 2011. Trust in typical and high-reliability contexts: Building and reacting to trust among firefighters. *The Academy of Management Journal,* 54(5): 999–1015.

Danish Ministry of Defence. 2011. *SAR DENMARK—Organisation: SAR DENMARK, VOL I.*

Danish Ministry of Defence. 2015. *SAR DENMARK—Operational Manual: SAR DENMARK, VOL. II.*

Deutsche Gesellschaft zur Rettung Schiffbrüchiger. 2012. *German Maritime Search and Rescue Service: Bases and tasks—Organisation and area of operation—Technology and equipment.* Deutsche Gesellschaft zur Rettung Schiffbrüchiger. Bremen.

Deutsche Gesellschaft zur Rettung Schiffbrüchiger, & Havariekommando. 2004. *Führung und Leitung zur Gefahrenabwehr auf See*: DV 100 See.

Elmshäuser, K. 2007. *Geschichte Bremens*. München: Beck.

Federal Bureau of Maritime Casualty Investigation. 2011. *Investigation report 143/11: Death of a crew member of the sailing yacht SPECIAL ONE on 30 April 2011 off Fehmarn* no. 143/11. Hamburg.

International Labour Organization. 1991. *C164—Health Protection and Medical Care (Seafarers) Convention*: No. 164.

International Maritime Organization. 2000a. *MSC/Circ.960 Medical Assistance at Sea*.

International Maritime Organization, 2000b. *Resolution A.894(21) International Aeronautical and Maritime Search and Rescue (IAMSAR) Manual*.

International Maritime Organization. 2005. *IMO Standard marine communication phrases*: *With pronunciation guide on CD-ROM*. London: IMO.

International Maritime Organization. 2006. *SAR convention: International convention on maritime search and rescue, 1979; as amended by resolutions MSC.70(69) and MSC.155(78); 2006 edition* (3rd ed.). London: IMO.

International Maritime Organization. 2011. *STCW—International convention on standards of training, certification and watchkeeping for seafarers: Including 2010 Manila ammendments; STCW Convention and STCW Code*. London: International Maritime Organization.

International Maritime Organization. 2014a. *ISM code: International safety management code with guidelines for its implementation; 2014 ed* (4th ed.). London: International Maritime Organization.

International Maritime Organization. 2014b. *SOLAS: Consolidated text of the International Convention for the Safety of Life at Sea, 1974, and its protocol of 1988: articles, annexes and certificates; incorporating all amendments in effect from 1 July 2014* (6th ed.). London: International Maritime Organization.

International Maritime Organization. 2015. *GMDSS manual: Global maritime distress and safety system* (8th ed.). London: International Maritime Organization.

International Maritime Organization; International Civil Aviation Organization. 2010a. *Mobile facilities: 2010 edition* (8th ed.). London: International Maritime Organization.

International Maritime Organization; International Civil Aviation Organization. 2010b. *Organization and management* (8th ed.). London: International Maritime Organization.

International Maritime Organization; International Civil Aviation Organization. 2013. *Mission co-ordination* (2013 ed.). London: International Maritime Organization.

Lipshitz, R., Klein, G., Orasanu, J., & Salas, E. 2001. A welcome dialogue—And the need to continue. *Journal of Behavioral Decision Making*, 14(5): 385–389.

Møller, A. *Guide to coordination of major SAR incidents at Sea: Mass rescue operations*; http://pre2016.balticsarex.org/images/documents/sarex2015/Guide_to_Coordination _of_major_SAR_incidents.pdf, September 19, 2016.

Russwurm, C. 1865. Ueber das Strandrecht in den Ostprovinzen.: Verlesen in der 249. Sitzung der Gesellschaft am 13. April 1860. In Gesellschaft für Geschichte und Alterthumskunde der Ostsee-Provinzen Russlands (Ed.), *Mittheilungen aus dem Gebiete der Geschichte Liv-, Ehst- und Kurlands*, 10: 3–24. Riga: Nicolai Kymmel's Buchhandlung.

United Nations. 1982. *United Nations convention on the law of the sea: UNCLOS*.

Weick, K. E., & Sutcliffe, K. M. 2007. *Managing the unexpected: Resilient performance in an age of uncertainty* (2nd ed.). San Francisco, CA: Jossey-Bass.

Weick, K. E., Sutcliffe, K. M., & Obstfeld, D. 1999. Organizing for high reliability: Processes of collective mindfulness. *Research in Organizational Behaviour*, 21: 81–123.

4 How Are Decisions in Complex Task Environments Actually Made?

Insights from Maritime Search and Rescue

4.1 A VARIETY OF OPEN QUESTIONS

As we outlined in Sections 2.3 and 2.4, we have a relatively clear understanding of the conditions under which intuitive and deliberate decision modes result in good choices. Intuition is particularly appropriate when decisions must be made quickly, holistic processing is important, and the link between cues and effects are stable—that is, when decision-makers are able to clearly and consistently link certain inputs to certain outcomes. Deliberation, in contrast, is particularly beneficial when outside information must be included in a decision, demanding analyses must be performed, and the biases of the intuitive mode are a concern. However, in practice these conditions are almost never clear-cut but, rather, tend to overlap. Therefore, these insights do not help us understand how decision-makers *actually* make decisions in complex task environments nor do they typically help us decide whether a decision should be or has been made using a particular decision mode. For example, in Chapter 1 we discussed the ill-fated decisions of the captain of a damaged search and rescue cruiser to first lead his crew to the forecastle for evacuation by helicopter and then back to the safe deckhouse when evacuation was ultimately not possible—a decision that eventually claimed his life when he did not make it back to the deckhouse safely in stormy weather. We will never know how the captain made this decision, but it is worthwhile to consider which approach would have been most appropriate. Should this decision have been made using a more intuitive or more deliberate mode? There were a variety of indications that making this decision intuitively was a good idea, such as the large number of available cues (sea state, state of his crew, wind, helicopter presence, influence of external conditions on helicopter operability, more cruisers on the way, damage to the ship, and so on). However, there were also good reasons to use a deliberate mode, such as the conscious calculation of risk and reward for the dangerous trip to the forecastle, avoidance of intuitive decision biases, and a lack of experience with comparable situations. Thus, it was by no means clear which decision mode the captain would use. Yet, the choice of decision mode would certainly have impacted the decision that was ultimately made. An intuitive decision could have led to substantial decision biases; attribute substitution (Kahneman & Frederick, 2002), for example, seems rather likely. It is conceivable that the captain, upon seeing the helicopter on scene, focused his decision-making on how to reach the forecastle, thus neglecting the broader and more relevant decisions of whether the risk of going to the forecastle was outweighed by the probability that the helicopter could successfully evacuate the crew given the weather and sea state conditions and whether the crew should instead stay in the deckhouse and wait for a safer evacuation route at a later time (see Figure 4.1).

In addition to understanding actual occurrences in action and deriving recommendations for how to deal with decision problems, the question of which decision mode decision-makers use in certain situations is important for the development of systems and trainings that help decision-makers reach effective decisions in complex task environments. Decision environments can be designed such that they are beneficial to the quality of decisions made using a particular decision mode. Therefore, designing decision environments such that they support effective decision-making is a means to increase the safety and quality of operations in complex task environments. We discuss this point comprehensively in Chapter 6 in the conclusion of the book. Finally, not knowing how decision-makers actually make decisions in

FIGURE 4.1 Hoisting maneuver in a mass rescue operation (picture from Baltic SAREX 2014).

complex task environments is also an important knowledge gap in academic research on decision-making in action that hampers progress in the development of our understanding of human decision-making in general and the "use factors" of intuition and deliberation in particular (Dane & Pratt, 2007).

Overall, answering the question of which decision mode decision-makers actually use in complex task environments is important, yet the answer will likely be "it depends." Considering what we have learned about deliberation in inherently important, judgmental decisions under situational complexity (Section 2.2), it seems likely that decision-makers will be reluctant to apply deliberation, as deliberation is taxing in terms of mental capacities and opportunity costs (Kurzban, Duckworth, Kable, & Myers, 2013). That is, when applying deliberation, decision-makers cannot do much else. In addition, naturalistic decision-making claims that proficient decision-makers, who are at the focus of our research, generally tend to make decisions using a strongly intuitive decision mode (see Section 2.3.3 as well as Kahneman & Klein, 2009; Klein, 1995). Therefore, it seems reasonable to assume that decision-makers will prefer intuition as a default strategy and deliberate only as much as necessary in a given decision situation based on experience, decision preference, and perceptions of the decision environment (see also Evans, 2006). Accordingly, the question is what triggers the perception that deliberation is required.

Considering the complexities of human decision-making in general and triggers of deliberation discussed previously, it appears likely that a variety of contingency conditions will affect individuals' choice of decision mode. First, it is clear that not all decision modes are equally feasible in all situations. Thus, how subjective perceptions of situational complexity affect decision modes is an important aspect. Some decisions, such as decisions under very high time pressure, need to be made very quickly—for example, a firefighter trapped in a burning building. Other situations clearly require an analytical approach—for example, mathematical optimization decisions. Yet, for the large majority of decisions of the type examined in our research (see Section 1.3), there is no such clear frame. Thus, we do not clearly understand if and how decision-makers in complex task environments adopt their choice of decision mode to different levels of situational complexity. Relatively low-complexity decisions, for example, could be solved with an intuitive decision mode, as they do not require the inclusion of outside information. However, they could also be solved using deliberate decision-making, as these decisions do not place a great deal of strain on the cognitive capacities of decision-makers. The same considerations can be made for decisions of a middle complexity level as well as those with high and very high complexity. For all levels of complexity, there are theoretical reasons arguing for the use of both decision-making modes. Thus, it remains a matter of exploration to identify the empirical relationship between situational complexity and decision mode. As stated earlier, there is reason to believe that intuitive decision-making should be dominant, but empirical evidence for the effect of varying degrees of complexity on decision mode is scarce. Therefore, this question remains largely unanswered.

Second, it remains to be seen whether decision-makers adjust their decision mode at all or, instead, consistently use their preferred decision mode. As discussed in Section 2.5, decision-makers have a preferred method of approaching difficult decisions that remains relatively stable over time. Some grab a pen and paper to make a list of pros and cons and then carefully compare the alternatives. Others prefer to look within themselves and go with what they feel in their heart is the right choice. Does this preference, which has been shown to be an influential and reproducible personality trait (Betsch, 2008), determine decision-making in complex task environments? Is this preference relevant at all when the situation becomes complex? Or is the effect somewhere in between?

Third, we learned in Sections 2.3 and 2.4 that experience plays a key role for the performance of both deliberate and intuitive decision-making (yet via different mechanisms). Although the relationship between experience and decision quality goes widely unchallenged, the relationship between decision mode and experience is less clear: Does experience affect how people make decisions? Previous work indicates that experience might cause lock-in effects, meaning that experienced individuals tend to become stuck in a particular approach and lose flexibility (Dane, 2010). This would imply that experienced decision-makers might stick more to their decision mode preference and also adjust their decision mode less to situational demands. Still, this would go against the notion that experienced decision-makers should have a better understanding of when they need to use more or less deliberation. It is currently unclear if such a lock-in effect is empirically relevant in complex task environments.

Figure 4.2 provides a schematic overview of these effects, which form the center of the following discussions.

In addition to the three major contingency conditions outlined in Figure 4.2, there are also a variety of other conditions that could potentially affect a professional's choice of decision mode. In the following, we briefly discuss some conditions relevant in the SAR context. Specifically, we examine the following:

> *Organizational background:* It is conceivable that organizational socialization affects how people make decisions (Pratt, Rockmann, & Kaufmann, 2006). Thus, we examined if and how a military or nonmilitary background

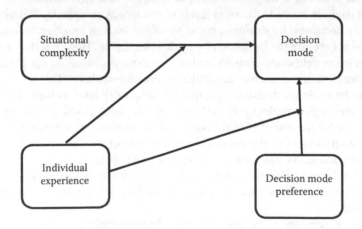

FIGURE 4.2 Theorized effects of situational complexity, individual experience, and decision mode preference on decision mode.

affects decision mode. This is an important question, as both military and nonmilitary professions are present in our empirical field.

Mindfulness: Mindfulness, "a state of consciousness in which attention is focused on present-moment phenomena occurring both externally and internally" (Dane, 2011, p. 1000), has been shown to be an important predictor of work performance and psychological well-being in complex environments. Although our studies do not primarily focus on mindfulness, we examined the degree to which mindfulness affects decision-making in complex tasks. Previous research indicates that decision-makers with strong mindfulness, that is, those who find it easy to focus on the present moment, apply more intuition (Laureiro-Martinez, 2014). In the studies described earlier, we expand on this claim. Mindfulness is an interesting construct, since it appears closely related to decision quality in complex task environments as studies on, for example, the performance of lawyers in court suggest (Dane, 2013).

Resistance to stress: The final concept we look into as a predictor of decision mode is resistance to stress, which captures a form of mental resilience in complex task environments. Stress resistance is, in this context, related to self-efficacy (Campbell-Sills & Stein, 2007), which is the belief in one's ability to master a specific situation (Bandura, 1977). This belief has a strong effect on both work performance in teams (Kozlowski & Ilgen, 2006) and individuals (Bandura, 1977) in challenging situations, whereas stress leads to a variety of negative behavioral outcomes that might affect the choice of decision modes. For example, people who are less resistant to stress tend to experience fatigue and, in the worst case, breakdown in demanding situations (Lazarus, 1993), which might impair their ability to apply deliberation. Coping with stress is, thus, important in demanding work contexts. It is conceivable that decision-makers displaying low stress resistance decide differently than those with high stress resistance.

In this chapter, we explore the question of how maritime SAR professionals make decisions during complex assignments. What is their preferred method of decision-making? What personal and situational factors affect SAR professionals' choice of a specific decision mode during decision-making? This helps us develop our understanding of decision-makers' choice of decision modes in complex task environments as an input for developing recommendations to further improve decision-making in complex task environments and narrow the knowledge gap in the academic literature.

4.2 EMPIRICAL BACKGROUND: THE CONCEPTS, STUDIES, AND MEASURES

The following insights rest on empirical studies that employed survey methodology in order to understand how decision-makers decide in real-life decision tasks, contingent on a variety of variables that are of interest to us (we will report video-based studies in Chapter 5). In Chapter 3, we explained the overall empirical approach taken in this book and presented the setting in which our studies take place. In the following, we will briefly outline the specific setup of the survey studies as far as they are relevant to understanding the following chapters.* We will present the general empirical approach as well as how we measured the respective contingency conditions that entered our empirical models. Thus, we will provide in-depth insights into different aspects of the studies in the following section. Readers not interested in this type of information may feel free to skip to Section 4.3.

The survey studies reported here were conducted and data collected at four different occasions (see Chapter 3 for more details): (1) a simulator exercise, conducted in a safe, in-house maritime SAR simulator in Germany, where professionals undergo routine training for complex tasks that seldom occur in real life, and (2) an evaluation of real-life SAR assignments that ran for 18 months and addressed SAR activities in German waters as well as two studies—Studies 3 and 4—conducted during large, open-sea exercises in Denmark, where decision-makers practiced complex tasks in a challenging open-water environment. As described in Chapter 3, in all three exercise events, which were hosted by maritime rescue service providers in Denmark and Germany, we cooperated with or were part of the steering team, ensuring exclusive access to participants as well as ensuring that the participants took the surveys seriously. For the real-life exercises, we also cooperated with leading personnel at the partnering organization, the German maritime SAR service provider. We used Studies 1 through 3 to develop the general insights we were interested in, whereas Study 4 served as a robustness test for the key relationships. For Study 4, we used somewhat different measures for key variables to establish whether the results obtained in the other studies might have been driven by our measurement approach. In addition, we looked at some specific variables in Studies 3 and 4 that were not part of our empirical approach in the other two studies, in particular regarding the effects of mindfulness and resistance to stress.

The tasks the participants had to undergo varied in terms of complexity, ranging from relatively routine search missions to highly complex mass evacuation tasks involving hundreds of personnel and dozens of vessels. Participants possessed different levels of experience as well as all other intrapersonal measures we examined. Still, all participants were well-trained professionals; thus, our samples do not include "novices." In total, SAR professionals from 11 different countries contributed to our sample, with the majority of the participants coming from the Baltic region: Denmark, Sweden, Finland, Germany, and Poland are the nationalities most strongly represented in the samples. All the following analyses were conducted while controlling for (i.e., statistically eliminating) the effect that national cultures might have had on the results. Table 4.1 provides more information on the sample.

We changed the measures to some degree between the events (see the following). However, the standard setup of the study remained constant. Before the exercise, we distributed a questionnaire to each participant capturing personality characteristics, such as experience, self-efficacy, and decision preference. We either collected this information before the first exercise began or, in the case of

---

* For a more academic perspective on the surveys with an extended methodology section, please refer to other publications of the authors, in particular Steigenberger, Fuchs, and Lübcke (2015).

TABLE 4.1

Empirical Studies and Number of Participants

Study	Empirical Field	Participants in Sample	Nationalities Represented in Sample	Decisions in Sample
1	Simulator exercise	57	Germany	306
2	Real-life data	23	Germany	63
3	Open-sea exercise	101	Denmark, Estonia, Finland, France, Germany, Latvia, Lithuania, Netherlands, Poland	387
4	Open-sea exercise	111	Denmark, Estonia, France, Germany, Latvia, Lithuania, Poland, Sweden	270

Study 2, before we distributed the questionnaires addressing real-life decisions. After we collected these initial questionnaires, we distributed the second questionnaire, which addressed decision properties, such as subjective decision complexity and the decision mode used. We instructed participants to fill out these questionnaires during their way back to port or immediately after reaching port following an assignment (Studies 2, 3, and 4) or directly after the simulator exercise was completed (Study 1). Respondents could report on three to six decisions made during an assignment; the maximum number varied between the studies. Both questionnaires were linked via an anonymized identifier code. This approach allowed us to link personality characteristics to situational properties without confusing the two levels (i.e., we avoid common method bias, see Podsakoff, 1986). Based on multilevel linear modeling techniques, we were then able to establish the effects of the conditions we were interested in.

The *decision mode* applied by a decision-maker was measured with a scale specifically developed for this study. To maximize the gain from having different studies available, we used two versions of this scale. One version captured deliberation with six items, including "We carefully compared the options we had" and "We had to give the situation serious thought." We used this version for Studies 1 through 3 and applied a plural form for the items after we learned that decision-makers typically consider the ship as the acting unit even though team decision-making does not play a role in this setting (however, see also the second case study in Chapter 5 on the role of groups in forming work routines). To test the robustness of this approach as well as the overall strength of our results, we employed a different scale for Study 4 based on the use of intuition with five measures, including "I made the decision that felt right for me" and "When making this decision, I relied on my feelings and emotions," using the singular form for item wording. As intuition and deliberation are opposite poles on a single spectrum in our model, we would expect inversely identical or at least strongly related results from both approaches if strong relationships were present.

Two dimensions of *subjective complexity* are of particular importance: the perceptibility of cues and the degree of uncertainty in a situation (Kahneman & Klein, 2009; Shiloh, Koren, & Zakay, 2001). However, complexity is field-specific, so that it must be understood on the basis of the specific tasks being examined. For maritime SAR, subjective complexity is based on a variety of environmental characteristics (Norrington, Quigley, Russell, & van der Meer, 2008)—in particular sea swell, wind, and visibility. These characteristics determine the visibility and perceptibility of cues as well as the decision-maker's ability to perceive cues, which might be substantially hampered under heavy sea conditions. Regarding uncertainty, familiarity with the local hazards is important, as is knowledge concerning the location of the vessel or person in distress. Then, the availability of resources affects cue perceptibility. If a vessel is, for example, under strain regarding crew size, it is much more difficult to perceive cues. We measured all dimensions of complexity with binary variables (favorable or unfavorable) and summed the unfavorable conditions to develop an understanding of aggregated situational complexity. Finally, there is a general element of surprise, which we captured directly with the question "While doing your job, did you and your colleagues find it easy to predict what would happen next?" This measure was also included in the complexity scale.

To test the robustness of this subjective complexity measure, we also developed an alternative complexity measure, which we used in Study 4. In this measure, we replaced the binary with a five-point scale, allowing more fine-grained answers, and also added a possibility for the respondents to indicate that a specific complexity dimension was not relevant for the focal decision. We also added two items capturing the degree to which information (i.e., cues) were readily available as well as how difficult it was to obtain necessary information.

For all other constructs that were of interest to us, we relied on straightforward and/or established measures. *Experience* was measured as years working at sea. Professionals in this field of work undergo routine training and are, in the case of an emergency, required to conduct or assist with SAR operations irrespective of whether they are SAR professionals or employed in other maritime professions. In addition, sea rescue service providers tend to hire personnel who already have maritime experience. Therefore, total years at sea is the appropriate measure for experience in this context.

Preference for intuitive decision-making was captured with an established scale developed by Scott and Bruce (1995). We employed the intuition-related subscale from the larger inventory of decision-related styles presented by these authors. Exemplary items include "I generally make decisions that feel right to me" and "When making decisions, I rely upon my instincts."

Military background was a binary variable capturing whether a decision-maker was currently working in a military organization.

Mindfulness is an established construct in psychology, meaning that well-evaluated scales are readily available. We employed the scale developed by Brown and Ryan (2003), which consists of 15 questions. Examples are "I rush through activities without being really attentive to them" and "I tend not to notice feelings of physical tension or discomfort until they really grab my attention."

Resistance to stress can be measured in different ways. We decided to use the scale developed by Campbell-Sills and Stein (2007), which covers mental resilience as the ability to cope with challenging and negative occurrences. Typical questions in this 10-item scale include "I am not easily discouraged by failure," "I can deal with whatever comes," and "I think of myself as a strong person." In maritime SAR, where physical and mental strain is commonplace, this measure is appropriate.

We also captured age and highest education as *biographical variables*. The sample consists almost entirely of men. We thus did not capture respondents' gender. In addition, to control for the effects of *time pressure*, we included time pressure as a specific measure in Study 4. We also controlled for the *type of decision* (command vs. operational) and captured a decision-maker's *satisfaction* with a decision using the question "How satisfied are you in retrospect with the decision?" Whereas the real-life nature of our approach with its incalculable and complex environments makes it almost impossible to directly capture decision quality, we employed this self-evaluation measure to obtain an approximation of decision quality, which, however, must be considered with care, as self-evaluations are substantially distinct from objective measures of decision quality, which are only obtainable in a laboratory setting.

We used most items in all four studies with the exception of mindfulness and stress resistance; those were only included in Studies 3 and 4. The inclusion of the control variables also varied to some degree between the studies. The following analyses are based on hierarchical linear modeling with decisions nested in decision-makers as a modeling data structure. See Steigenberger, Fuchs, and Lübcke (2015) for the technical details of the analyses.

4.3 RESULTS

In the following, we will present the results, focusing on the relationships between decision mode and the three variables that we are primarily interested in: decision preference, situational complexity, and experience. We will also briefly examine predictors of decision mode preference and drivers of satisfaction with a decision. Afterward, we will examine the other variables in the studies, sum up the results, and draw conclusions. The following results are drawn from all studies. However, we will emphasize the results of Studies 3 and 4, which provide the richest insights due to having the highest numbers of responses as well as because we feel that these studies offered the ideal combination of real-life validity and experimental control (see Sections 3.2 and 4.2). We will also explicitly discuss the instances in which the results differed between studies.

4.3.1 General Distribution of Decision Modes

We will begin our analyses by examining how decision-makers actually decide in complex task environments. That is, we will examine the decision mode used.

GENERAL DISTRIBUTION OF DECISION MODES

- Roughly normally distributed
- No effect of military background, age, education, or experience

Overall, we see that intuition does not dominate as a decision mode. Instead, there is an approximately normal (i.e., bell-shaped) distribution of decision modes used over the full spectrum of the scale with the mean of the distribution slightly on the intuitive side and relatively large variance (standard deviation). However, we observe some differences between the studies. Decisions in the indoor simulator environment (Study 1) were relatively strongly on the intuitive side, whereas decisions in the open-sea exercise (Study 3) were, on average, the most deliberate ones. The decision modes used in the real-life data were somewhere in the middle.*

This variation between the studies can be explained in two ways. First, the average complexity of the tasks in which the decisions were embedded varied somewhat between the studies. Whereas the tasks for the routine trainings in the simulator environment were, on average, comparatively less challenging, the open-water exercises captured in Studies 3 and 4 were specifically designed to provide a substantial challenge. Real-life occurrences (Study 2) are in the middle—that is, more challenging due to being outside of the safe zone that the simulator environment provides but not as challenging as the highly difficult tasks in the high-profile open-sea exercise. As we see in the following, decision-makers tend to approach low-complexity decisions with an intuitive decision mode.

Second, a characteristic of the learning structure in the large-scale exercises might help us understand the differences between the studies. The large-scale exercises were built strongly around debriefings of the decisions made, for which the debriefings involved explaining decisions in a foreign language (due to the international character of the exercises) to colleagues with whom the decision-makers do not typically work. This subtly encourages decision-makers to use more deliberation. Deliberate modes involve a conscious selection of rules. Thus, it is much easier to discuss a decision made with a strongly deliberate mode compared to decisions made with a strongly intuitive mode in which decision rules remain largely on the subconscious level. As discussed in Section 2.4, it is one of the advantages of deliberation that the decision rules used are typically consciously perceived and can thus be reproduced with relative ease. This is also an explanation for the relatively strong level of deliberation in Studies 3 and 4. The simulator study (Study 1) took place in a more intimate and less challenging environment, so the effect of having to explain decisions to an unfamiliar audience is less pronounced there. These findings are the first indication that the setting in which a decision takes place influences the choice of decision mode.

The most important findings are, however, that we did not observe one dominant decision mode as well as that extreme cases (purely intuitive and purely deliberate decision-making) are rather rare. Instead, some aspects of intuitive and deliberate

---

* As we used a different scale to capture decision mode in Study 4 (see Section 4.2), it is not possible to directly compare Study 4 with the other studies on this variable.

decision-making enter into most decisions. The question we address in the following is to explain which conditions shape the degree to which deliberation is used or, to frame it differently, the degree to which decision-makers trust their intuitions.

4.3.2 DECISION MODE PREFERENCE

The next major question is whether decision mode preference determines the decision mode actually used in complex tasks. Decision mode preference is a habituated response pattern to decision situations (Scott & Bruce, 1995). Through continual experience making decisions using a specific mode, individuals develop a personal preference for a particular decision mode, which lies somewhere on the continuum between full intuition and full deliberation. This preference is a stable personality construct (Epstein, Pacini, Denes-Raj, & Heier, 1996; Evans, 2010; Evans & Stanovich, 2013)—that is, it changes little over time. Decision mode preference describes a tendency for decision-makers to use a specific decision mode largely irrespective of the circumstances. Currently, we do not know to which degree decision mode preference affects or even determines the choice of decision modes in complex task environments. For the initial considerations, we assume that although decision mode preference is presumably influential, it is unlikely to predetermine decision-making (see Section 2.5). To our knowledge, decision mode preference has previously only been addressed in experimental, laboratory research (cf. Betsch, 2008; Pretz, 2008). Our studies, therefore, add substantially new evidence. For the training of decision-makers and the construction of favorable decision environments (Chapter 6), it is important to have an understanding of whether or not decision-makers actually switch to decision modes they are not perfectly comfortable with.

First, we began by examining the distribution of the decision mode preferences to establish whether the respondents in our setting shared a preference for a particular mode. We also examined the effect that professional background (military vs. nonmilitary), stress resistance, mindfulness, and demographic variables had an influence on which decision mode a decision-maker prefers.

Across the whole sample, we found an approximately normal (i.e., bell-shaped) distribution of decision modes with the highest frequency around the scale mean. This means that decision-makers in our sample show, on average, a strong preference for neither intuition nor deliberation, and only a few respondents reported extreme values. It is therefore safe to conclude that there is not one particular type of decision-maker overrepresented in the field of maritime SAR. We also found that military background had no influence. That is, military and nonmilitary personnel did not differ in terms of their preferred approach to the choice of decision modes. This finding is somewhat surprising, as we know that employees from the same professional background tend to share many work characteristics (Chreim, Williams, & Hinings, 2007). This indicates that professional background does not strongly differentiate employees with respect to decision mode preference.

The same is true for the other biographical variables. Education and age have only marginal effects: Younger and less educated personnel have a slightly

THE ORIGIN OF DECISION MODE PREFERENCE

- Stress resistance is associated with a preference for intuition.
- Mindfulness is associated with a preference for deliberation.
- No effect of military background, age, education, and experience.

stronger tendency toward intuition. However, this effect is almost negligible in strength; and there is no effect of experience. However, we found a quite substantial relationship between stress resistance and decision mode preference. That is, stress-resistant personnel display a substantially stronger preference for intuitive decision-making compared to employees lower in stress resistance. This finding is interesting yet not surprising. As discussed in Section 2.4.1 and earlier, a substantial disadvantage of the intuitive decision mode is that decision-makers are typically unaware of why they arrived at a certain decision, making it difficult to justify, in retrospect, a decision made with an intuitive mode. For example, being able to consistently stand up in a debriefing saying that a decision was taken based on gut feeling certainly requires the form of mental resilience that is captured here with items such as "I think of myself as a strong person" or "I can deal with whatever comes." We will discuss the implications of this finding in Chapter 6.

Mindfulness has the opposite effect of stress resistance: Highly mindful personnel prefer deliberation more as compared to less mindful personnel. This finding is somewhat surprising as mindfulness tends to be associated with holistic thinking, which is a property of an intuitive decision mode (Dane, 2011). Yet, in our case, we see that highly mindful decision-makers tend to have a preference for deliberate analysis. This finding can be understood with the characteristics of mindfulness and deliberation. We see that persons who are strongly focused on the current moment find it less taxing to analyze situations deliberately, as conscious and deliberate cue processing comes easier to them; and they are accordingly better able to select important cues and thus decrease the unwelcome cognitive load of deliberate processing.

All these effects are independent of each other; and there are no relevant interactions between them. As a next step, let us look at the relationship between decision mode preference and actual decision-making. We would expect that decision-makers with a preference for intuition use more intuition in their actual decision-making and vice versa. Not surprisingly, we found this effect in most studies. We are thus able to support previous laboratory findings based on our data obtained in a more realistic setting. More interestingly, we saw that the strength of this effect is quite limited. In particular, it does not overrule the effects of the other variables we examined, for example, situational complexity. This means that although decision mode preference is of some importance, it does not dominate the choice of decision modes in actual decision-making. Decision mode preference instills a tendency for choosing a certain decision mode but not more.

Looking at interactions with other variables of interest, we see that decision mode preference interacts with experience such that less experienced decision-makers tend to follow their decision preference more strongly than more experienced decision-makers. Experienced decision-makers do not follow their decision mode preference, which is unrelated to a specific task or even work domain, less strongly. We can thus speculate that experienced decision-makers develop field-specific decision routines, which are driven by the task and experience with the task and the field instead of individual preferences. This finding might be seen as an indication that experience helps decision-makers choose the appropriate decision mode in a complex situation. However, we will see later in Section 4.3.4 that this interpretation has substantial limitations.

THE EFFECTS OF DECISION MODE PREFERENCE ON THE CHOICE OF DECISION MODE

- Preference for intuition leads to more intuitive decision-making.
- Decision mode preference does not negate the effects of other variables, such as situational complexity; thus, it does not determine the choice of decision mode.
- Highly experienced decision-makers follow their decision preference less than less experienced decision-makers.

4.3.3 SUBJECTIVE SITUATIONAL COMPLEXITY

Having established that decision mode preference does not determine the choice of decision modes in complex task environments, we can now turn to the most interesting variable: situational complexity. In our conceptualization, situational complexity is subjective (see Section 1.3). The same decision might therefore be perceived as highly complex by one decision-maker and not very complex by another. We are thus not interested in objective decision difficulty but, rather, in how a specific decision-maker experienced the complexity of a specific decision in the dimensions outlined above (see Section 4.2).

First, we looked at the overall relationship between subjective situational complexity and the choice of decision modes. Importantly, we found that this relationship is not linear but, rather, inverse U-shaped. Decisions on the low end of the complexity scale are largely made with an intuitive decision mode. Based on our expectation that low-complexity decisions are often routine and, thus, do not typically require or trigger conscious deliberation, this observation came as no surprise.

With an increase in subjective situational complexity, however, intuition is used less and less—that is, growing subjective situational complexity increases decision-makers' use of deliberation. Situational complexity was defined as a combination of unfavorable weather conditions, missing information pertaining to the decision situation, a lack of necessary resources, and unpredictability regarding what would happen next (see Section 4.2). This combination of factors apparently made it increasingly difficult for decision-makers—who as SAR professionals are arguably proficient in their field—to apply intuitive decision-making. This indicates that the applicability of routine-based intuition is limited as well as that decision-makers subsequently experience a need to develop a more abstract understanding of a situation and elaborate more thoroughly under growing complexity.

This finding is fairly surprising, as it directly contradicts predictions of naturalistic decision-making research, which claims that under conditions of higher complexity competent decision-makers should strongly rely on intuition in the form of recognition-primed decision-making. This means that they should rely on subconscious cue evaluation and only occasionally complement this decision mode with mental simulation as a limited form of deliberate decision-making (Klein, 1995; Nemeth & Klein, 2011). Yet this is clearly not what we found. Instead, we see a strong increase in the use of deliberation, which continues for a large part of the complexity spectrum. We will discuss this further in Section 4.4.

THE DIRECT RELATIONSHIP BETWEEN SITUATIONAL COMPLEXITY AND DECISION MODE

- Low-complexity decisions are, to a large degree, intuitive decisions.
- Increases in complexity trigger more and more deliberation.
- At very high complexity levels, there is a turning point; and decision-makers turn back to somewhat more intuitive decision-making.

Although the tendency to increase deliberation with an increase in complexity is strong and pronounced, there is a turning point. At high levels of complexity, decision-makers return to a more intuitive decision mode. The point at which this occurs indicates the limits of deliberative processing. Processing information consciously is demanding, and especially when the situation is unclear, information is difficult to obtain, or the environmental conditions are rough, the capacity for deliberate processing will at some point be overwhelmed (Dijksterhuis, 2004; Dijkstra, Pligt, & Kleef, 2013). The ability to apply more deliberation will be exhausted, as the decision-maker will at some point be unable to process information consciously and deliberately. Intuition with its holistic, fast, and automated information processing does not suffer from this problem and is, therefore, generally applicable throughout the entire spectrum of complexity (Kahneman & Klein, 2009). Under high complexity, the disadvantages of deliberate decision-making thus eventually take predominance. However, although we find an indication that decision-makers return to some degree to intuition, the downturn is relatively small; only at the end of the spectrum do we see that intuitive decision-making once again takes predominance. Figure 4.3 shows, as an example, the relationship between decision mode and situation complexity as observed in Study 3.

Figure 4.3 plots the inverse U-shaped relationship, in which the value "0" on the y-axis refers to the sample mean. We see that, over a relatively large range of complexity values, on the low end of the complexity spectrum decision-makers use more intuition than the sample average. The use of deliberation then increases steadily up to a maximum; and it is only at the very end of the complexity spectrum that decisions return to a below-average level of deliberation.

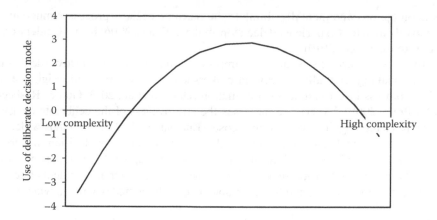

FIGURE 4.3 Relationship between situational complexity and use of deliberation in Study 3. Plot based on the tool provided by Jeremy Dawson: http://www.jeremydawson.co.uk /slopes.htm.

Looking further, we see that the relationship between subjective complexity and decision mode is neither dependent on nor does it interact with stress resistance. In some studies, the relationship between complexity and decision mode is somewhat affected by the mindfulness of the decision-maker, which we will discuss Section 4.3.4. There is also an inconsistent effect of decision mode preference, which stems from the fact that those with an intuitive decision mode preference use a more intuitive decision mode; accordingly, the shift toward deliberation is weaker for those with a strong preference for intuition as would be expected. More critically, we see a strong and relevant interdependence between subjective complexity, experience, and decision mode, which we will discuss in the following.

4.3.4 EXPERIENCE

Experience is an important force in decision-making in complex task environments as decision-makers develop domain-related knowledge (Gioia & Poole, 1984; Lieberman, 2000) when they are repeatedly exposed to comparable situations. Although not all work contexts are equally suited as a learning environment (we will discuss "wicked" learning environments in Chapter 6), as a general rule we see that experience strongly benefits both decision modes. For intuition, the relevant domain-related knowledge provides the grounds for "mature" intuition (Baylor, 2001). That is, the intuitive decision mode holistically relates perceived cues to previous experiences and thus draws on prestored patterns and tacit insights— knowledge that would not be readily available based on conscious thought (Dane & Pratt, 2007). Based on this domain-specific knowledge, intuitive decision modes enable highly experienced decision-makers to link sensory inputs with potential decision options and mentally simulate the outcome of these processes much more effectively than less experienced decision-makers (Evans, 2006; Klein, Calderwood, & Clinton-Cirocco, 2010).

However, experience is also an important contributor to effective deliberate decision-making, as it allows decision-makers to quickly classify and weigh cues as well as process larger chunks of information (Dijkstra et al., 2013; Goll & Rasheed, 2005; Pretz, 2008). Experience increases the effectiveness of the deliberate decision mode, reducing cognitive opportunity costs (Kurzban et al., 2013) and thus widening the bottleneck that deliberate decision-making tends to impose on decision-makers in a complex task environment. This, in turn, increases the applicability of the deliberate mode. Experienced decision-makers should therefore be better at both deliberate and intuitive decision-making and be better able to apply both modes across a wider range of complexities.

Having established the overall importance of experience for a decision-maker's decision performance, let us have a closer look at the effects experience might have on cognition and the subsequent choice of decision modes. On the one hand, it seems likely that experienced decision-makers have a more fine-tuned understanding of the needs of a particular situation based on their domain-specific knowledge, which would imply that they would be able and willing to adjust their decision mode more to situational demands (Betsch & Glöckner, 2010). Experienced decision-makers might find it easier to decide when they require additional (outside) information or when they should apply complex rules as well as when the cues perceived in a particular situation are sufficient to make a good and fast decision. They should be less likely to fall prey to oversimplification or attribute substitution (Kahneman & Frederick, 2002), which might mean that less experienced decision-makers fail to recognize the required decision in a particular situation. Experienced decision-makers might also have a better understanding of when their cognitive capacities are going to be overwhelmed, so they might go back to intuitive decision-making earlier than less experienced decision-makers when complexity becomes critically high (Plessner, Betsch, & Betsch, 2008). This reasoning would imply that experienced decision-makers should fine-tune their decision mode more to situational demands—that is, situational complexity.

On the other hand, there is also reason to believe that experience might have quite the opposite effect. Most importantly, there has been some discussion on possible lock-in or "entrenchment" effects (Dane, 2010) caused by strong experience, which means that highly experienced decision-makers might develop strong decision-making routines they find difficult to change, irrespective of situational demands. Experienced decision-makers might become "entrenched" in that they always approach decisions in the same manner and adapt less to situational demands than less experienced decision-makers. There has been evidence that experts have a tendency to become fixated on thoughts that are triggered by their prior experience (Bilalić, McLeod, & Gobet, 2010; Dane, 2010) and thus may fail to consider other information and alternative approaches.

Experts' decision-making quality may also be hampered due to overconfidence if highly experienced decision-makers apply their knowledge in the wrong domain (Kahneman & Klein, 2009). Hecht and Proffitt (1995), for example, asked individuals to assess the liquid surface orientation of tilted containers and compared the performance of bartenders and waitresses with participants from other professional fields. Liquid surfaces naturally remain parallel to the ground independent of the angle at which their container is held. Waitresses and bartenders who work with beverages on a daily basis, and thus possess a great deal of experience with the behavior of liquids in containers that shift their position, are expected to be more aware of this fact than participants from other professional fields. Surprisingly, Hecht and Proffitt found that the performance of waitresses and bartenders, in terms of determining the correct angle of liquids in a tilted container, was worse than that of inexperienced participants. The researchers attribute the poor performance of waitresses and bartenders to their preoccupation with spilled liquids. In the context of our studies, this might imply that experienced decision-makers fail to recognize changes in the environment that might actually be important for the choice of decision modes. Comparable evidence comes from Lewandowsky and Kirsner (2000). Overall, there is evidence that experience has a variety of effects on the choice of decision modes and subsequent decision quality, and it is not clear how experience will affect the choice of decision modes.

In light of these theoretical considerations, we now discuss our empirical results. When we began empirically studying the effects of experience in our four studies, it became clear that experience had no direct effect on the choice of decision modes—that is, experienced decision-makers were not more likely than less experienced decision-makers to apply intuition or deliberation. Considering the heterogeneous effects that experience has, this is not a surprise despite the fact that these findings, to some degree, contradict the predictions of naturalistic decision-making research—specifically, that highly experienced decision-makers would rely on a strongly intuitive mode based on subconscious pattern-matching and semiconscious mental simulation (Klein et al., 2010; Klein, 1995).

As experience has no direct effect on the choice of decision modes, it is necessary to look closely at how experience affects the relationships between the other variables of interest change. First, we see that experienced decision-makers follow their decision mode preference to a lesser extent than less experienced decision-makers. We already discussed this finding in Section 4.3.1. More interesting is the interaction between experience and situational complexity regarding the choice of decision mode. Our findings suggest that the inverse U-shaped relationship, outlined in Figure 4.3, is much more pronounced for inexperienced as compared to experienced decision-makers.

Inexperienced decision-makers address low-complexity decisions with a routine-based decision mode and then begin to

DIRECT AND INDIRECT RELATIONSHIPS BETWEEN EXPERIENCE AND DECISION MODE

- Experience has no direct effect on the choice of decision modes.
- Experienced decision-makers adjust their decision modes less based on changes in subjective situational complexity than less experienced decision-makers.

search for additional information and utilize explicit deliberation as complexity increases. At some point, the decision situation is such that the downsides of deliberation, in particular high cognitive load, become overwhelming, at which point comparatively less experienced decision-makers return to intuition (see Section 4.3.2).

In contrast, highly experienced decision-makers do not show the U-shaped pattern displayed in Figure 4.3. Instead, for this subgroup of decision-makers, we find a linear relationship between subjective decision complexity and decision mode. Figure 4.4 plots this relationship for Study 3. Here, we see that the adjustment in situational complexity is not statistically significantly different from zero for the group of highly experienced decision-makers. Instead, those decision-makers approach decisions relatively deliberately and stick to this approach over the full spectrum of subjective complexity. Less experienced decision-makers, in contrast, display a pronounced version of the inverse U-shaped relationship we saw in Figure 4.3. However, the specific characteristics of the linear relationship differ somewhat between the studies. In Study 4, there is a linear trend to more intuition for highly experienced decision-makers and the same U-shape relationship for the less experienced decision-makers. However, the trend for highly experienced decision-makers is quite weak, whereas the overall level of deliberation used by experienced decision-makers is somewhat lower than the level we found in Study 3. Irrespective of the specific form of the linear relationship, it is clear that highly experienced decision-makers adjust their decision mode much less to changes in subjective complexity than less experienced decision-makers.

This might indicate that experienced decision-makers do not perceive increases in situational complexity to be pressing enough to warrant a change in decision mode. Yet, the settings examined in our studies, in particular in Studies 3 and 4, were

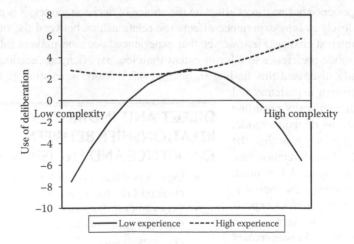

FIGURE 4.4 Relationship between situational complexity, experience, and use of deliberation in Study 3. Plot based on the tool provided by Jeremy Dawson: http://www.jeremydawson .co.uk/slopes.htm.

explicitly designed to include highly complex situations. Therefore, this explanation is not fully satisfactory. Instead, we might see empirical support for the notion that experienced decision-makers lose flexibility in their choice of decision modes—that is, they find it harder to adapt to changing environmental demands (cf. Dane, 2010). We will discuss this finding in more detail in Section 4.4.

4.3.5 OTHER VARIABLES OF INTEREST

We now examine the other variables of interest, beginning with *mindfulness*. Overall, the effect of mindfulness is rather inconsistent. In particular, in Study 4, mindfulness has no effect at all on the choice of decision mode. In Study 3, it led to some tendency toward more intuition, taking into account the fact that highly mindful decision-makers also have a slight tendency to prefer deliberation. Therefore, these findings are inconclusive; and it appears that there is no strong effect of mindfulness on the choice of decision modes, at least in our sample and studies. However, mindfulness indirectly affects the choice of decision mode in some ways. First, mindfulness amplifies the relationship between decision mode preference, experience, and actual decision mode. We saw before that experienced decision-makers follow their decision mode preference less than less experienced decision-makers. Adding mindfulness to the picture, we find that decision-makers who are low on both mindfulness and experience follow their decision mode preference most strongly. This provides support for the interpretation that following one's decision mode can be understood as being distracted from, and thus not responding sensibly to, actual situational demands.

> **DIRECT AND INDIRECT RELATIONSHIPS BETWEEN MINDFULNESS AND DECISION MODE**
>
> - Inconsistent evidence regarding direct effects of mindfulness on decision mode.
> - Those with low mindfulness and low experience follow their decision mode preference most strongly.

Mindfulness, as the ability to be "present" in the current moment, accordingly mitigates this problem. If persons are inexperienced as well as not very mindful, they simply go with their decision preference. This supports the notion that mindfulness, although it does not directly affect the choice of decision mode, might help decision-makers obtain a better understanding of the situational demands and adjust their decision mode accordingly. However, this effect is rather limited. Relatedly, was also saw that in Study 3 mindfulness increased the relationship between subjective situational complexity and decision mode such that highly mindful decision-makers adopted their decision mode more strongly to changes in subjective situational complexity, which also strengthens the argument above. However, this effect is not visible in Study 4.

Somewhat surprisingly, we found no effects related to *resistance to stress*. We saw that resistance to stress affected the general preference for a specific decision mode. Yet, there is no relationship between resistance to stress and the

DIRECT AND INDIRECT RELATIONSHIPS BETWEEN STRESS RESISTANCE AND PERCEIVED TIME PRESSURE ON DECISION MODE

• There are no effects of stress resistance as well as time pressure on the choice of decision mode.

actual choice of decision modes, neither directly nor indirectly.

To rule out alternative explanations for our results, we directly tested the effect of perceived *time pressure* on decision mode in Study 4. Time pressure had no effect by itself. In addition, the relationship discussed above for Study 4 holds when we include time pressure as an explicit condition in the statistical models, strengthening our claim that, at least in this specific setting, time pressure is not an immediate concern for decision-makers, although the time for each decision is, of course, generally limited.

4.3.6 SATISFACTION WITH A DECISION

Finally, we examine decision-makers' *satisfaction with a decision*. Although not a full-fledged measure of decision performance, this variable provides an impression of the results of a retrospective self-evaluation and, with that, insights on the outcome of a decision. First, we see that, although decision-makers were overall relatively satisfied with their decisions, there is sufficient variance in this variable to use it meaningfully in follow-up analyses. This is not the case in the real-life data (Study 2), however, where decision-makers rated almost all decisions very positively. This reflects the notion that real-life decisions tend to be relatively uncomplicated for seasoned individuals in most cases (see Chapter 3 on issues with the real-life data). We therefore focus the following analyses on the other three studies, in which decision difficulty was overall higher, as the scenarios addressed had a different quality than what decision-makers typically encounter in their day-to-day assignments.

First, we found that decision mode alone does not affect satisfaction with a decision. This strengthens our confidence in this measure as a proxy for objective decision quality. We would expect that decision-makers would be more satisfied with intuitive decisions if the measure reflects only a purely subjective evaluation

PREDICTORS OF SATISFACTION WITH A DECISION

• The decision mode does not directly affect satisfaction with a decision.
• Decision-makers are less satisfied with complex as compared to less complex decisions. This relationship is stronger for decision-makers with strong experience.
• The relationship between experience and satisfaction with a decision is generally positive.
• Decision-makers with strong mindfulness are more satisfied upon deviating from their decision mode preference than less mindful decision-makers.

of a decision, as intuitive decisions tend to create feelings of certainty (Dane & Pratt, 2007). As we do not observe such a relationship, we have a good indication that the satisfaction measure approximates objective decision performance and does not merely reflect cognitive mechanisms of the chosen decision mode.

Looking further, we see that decision-makers are more satisfied with decisions for which they applied more intuition than their decision mode preference suggests and less satisfied with decisions for which they applied more deliberation than their decision mode preference suggests—while controlling for the complexity of the decision. This reflects to some degree the notion that intuition tends to come with feelings of certainty and is less taxing regarding cognitive demand (Dane & Pratt, 2007). However, this observation is not duplicated in Study 4. Thus, the strength and stability of this affect is somewhat unclear, which might, as stated above, be seen as an indication that the satisfaction with a decision measure reflects objective decision quality more than cognitive mechanisms linked to decision modes. Relatedly, we also see that decision-makers with strong mindfulness respond more positively to a deviation from their decision preference in a specific decision situation than decision-makers with low mindfulness. This is true for deviations toward more deliberation as well as toward more intuition. This finding reflects the ability of highly mindful decision-makers to effectively respond to situational demands as measured by changes in subjective situational complexity. Highly mindful decision-makers have a better understanding of when they need to deviate from their decision preference.

In terms of situational characteristics, subjective situational complexity consistently has a negative effect on satisfaction with a decision. That is, decision-makers were less satisfied in retrospect with highly complex as compared to less complex decisions. This is the expected relationship, reflecting the notion that high complexity implies a larger probability of missing or misinterpreting important cues, applying scarce resources wrongly, or judging uncertain outcomes incorrectly. Complexity should relate negatively to decision quality, irrespective of other variables (see also Sections 1.3 and 4.3.2). This is what we see in the data.

Examining the effect of experience on satisfaction with a decision, we again obtain slightly inconclusive results. In Studies 1 and 3, the effect of experience on satisfaction with a decision is positive and significant, as we would expect based on the discussions in Sections 2.3.3 and 2.4.3. However, this relationship is not significant in Study 4. Thus, it appears that experience has some effect on satisfaction with a decision, yet the effect is not so strong that it would reproduce consistently. To some degree this goes against the notion that experience aids deliberate as well as intuitive decision-making (see Section 4.3.3), for which there are two possible explanations.

The first explanation refers to the lock-in effect of experience that we discussed previously (Section 4.3.3); as experienced decision-makers adjust less to situational complexity, they might apply decision modes that are less well suited for a particular medium or highly complex decision, leading to poorer decisions in these complexity ranges than less experienced decision-makers. This effect would then to some degree counterweigh the positive effects of experience, in particular in generally highly complex tasks, such as the ones under observation in Study 4 (mass evacuation with a large number of actors involved). Support for this explanation is also found in Studies 1 and 3, looking at the effect of experience and complexity simultaneously (i.e., the interaction effects). We see here that for those with low experience—that is, those decision-makers who display the U-shaped adjustment of their decision mode to changes in subjective situational complexity—complexity does not significantly influence satisfaction with a decision. That is, inexperienced decision-makers show the same level of satisfaction, irrespective of complexity. Experienced decision-makers, on the other hand, are substantially less satisfied with their decisions in highly complex situations than with their decisions in less complex situations.

An alternative explanation is that experienced decision-makers, having a much wider range of decision options at their disposal, might be more critical with regard to their own decisions than less experienced decision-makers, who do not have such a broad understanding of which alternative options they might have had and who might not fully understand the consequences of a particular decision. This might also explain why highly experienced decision-makers rate their decisions under high complexity, in particular, especially critically.

Our measure for satisfaction with a decision has limits. Thus, we cannot clearly answer the question of which of these two explanations is more important. It seems likely that both mechanisms are in place and that the lack of adjustment to situational complexity, in particular, limits to some degree the beneficial effects of experience regarding decision quality.

In addition, we found that the specific character of a decision (command decision or operational decision) as well as decision mode preference have no effect on a decision-makers' satisfaction with a decision.

4.4   SUMMARY AND DISCUSSION OF THE SURVEY RESULTS

After conducting four studies in different settings, we are prepared to provide a preliminary answer to the questions of which decision mode decision-makers in complex task environments choose as well as how this choice affects decision outcomes. In the following, we will discuss the key findings.

Not surprisingly, there is no single answer to the question of how decision-makers approach complex decisions. Instead, a variety of conditions affect this relationship, many of which are somewhat inconsistent and, therefore, might not be generalizable or sufficiently strong to replicate reliably. Still, we obtained a set of core observations that replicated over different task contexts and also somewhat different measures for the core constructs. These relationships form our core insights and refer to the key characteristics we were interested in: subjective situational complexity, experience, and decision mode.

Looking at the main conditions of interest as outlined at the beginning of this chapter—that is, the relationship between decision mode, situational complexity, individual experience, and decision preference—we can answer the question posed in Figure 4.2 as follows:

First, decision-makers have a tendency to follow their decision mode preference in complex task environments, but this preference does not, by any means, determine which decision mode a decision-maker will eventually choose. This finding supports the initial notion of our book that environmental conditions, specifically situational complexity, affect how proficient decision-makers approach decisions in complex task environments. This finding is per se not particularly surprising. Yet it is one of the first times that the effect of decision-mode preference has been tested and observed outside of a laboratory setting (cf. Betsch, 2008; Pretz, 2008). In addition, it helps us disentangle methodologically the effects of decision mode preference and situational conditions. The following relationships, in particular those between situational complexity and decision mode, hold even if decision mode preferences are taken into account. Hence, we can be sure that the effect of situational complexity on decision mode is not a spurious effect caused by unobserved decision mode preference.

Perhaps the most critical and interesting insight provided by the studies concerns the relationship between decision mode and situational complexity. Here, we began with the notion of naturalistic decision-making that proficient decision-makers approach highly complex decisions with a recognition-primed, largely intuitive approach (Kahneman & Klein, 2009; Klein, 1995). We did not find such a relationship. Instead, we see that the choice of decision modes in relation to subjective situational complexity has an (inverse) U-shaped form: Decision-makers approach decisions they perceive as not very complex with an intuitive mode, then increase the degree of deliberation put into a decision with increasing complexity up to a point at which decisions tend to become more intuitive again with an additional increase in complexity. This pattern aligns well with ideas of metacognition, developed in conceptual and laboratory settings (Alter, Oppenheimer, Epley, & Eyre, 2007; Thompson, Prowse Turner, & Pennycook, 2011): Intuition appears to be the default mode for decisions for which no particular environmental trigger for deliberation exists. When cues become more difficult to perceive or use, resources become scarce, and uncertainty increases, decision-makers experience such a trigger of deliberation. As a result, they place more and more deliberate effort into a decision to the point at which complexity is so high that the deliberate decision mode becomes overwhelmed.

Considering the strengths and weaknesses of both deliberate and intuitive decision-making (see Sections 2.3.2 and 2.4.2), this pattern is very likely effective. Under low complexity, routine-based intuition is likely to produce good decision outcomes (Baylor, 2001), whereas the need for outside information or the conscious selection of rules is typically not an issue. Routine decision-making based on intuition frees up cognitive resources for more demanding tasks and is therefore appropriate. These are the decisions that decision-makers tend to not even perceive as a decision at all but, instead, complete automatically—that is, subconsciously. Uncertainty, resource scarcity, and more difficult circumstances in terms of weather or sea state then imply that the benefits of deliberate decision-making become more important: Decision-makers need to acquire additional information, analytically compare options, and form opinions on probabilities. In short, they must use more deliberation. As complexity increases and these benefits become increasingly more important, the strain on the working memory capacities also increases. At some point, the bottleneck gets too narrow and decision-makers reach a point at which they cannot place more deliberation into a specific decision or when the costs to do so, in terms of an inability to do something else, are perceived as too large. At this point, it again becomes effective to trust the intuitive mode to produce solid solutions based on recognition and pattern-matching (Klein, 1995). Thus, it appears that the (inverse) U-shaped relationship we observed reflects a decision-making approach that we would expect to be effective based on our previous understanding of the benefits and downsides of intuitive and deliberate decision-making. Still, the degree to which deliberation is and intuition is not used is somewhat surprising, as we would have expected to see intuition used over a broader range of the spectrum. In particular, in the studies in which complexity of the broader task was high or very high (i.e., in Studies 3 and 4), the overall approach to decision-making was substantially on the deliberate side. It seems likely that structural aspects play a role here, which we discuss more extensively in the concluding Chapter 6. On the other hand, this also represents strong evidence that the importance of intuition in decision-making might have been overrated in our previous understanding of decision-making in complex task environments, especially for more complex decision situations.

The relationship between situational complexity and decision mode is inverse U-shaped. Yet the reversal to an intuitive decision mode under exceedingly high complexity is relatively small, though more pronounced in Study 4 than in Study 3 (comparing only the two open-water exercises). It seems that the complexity of decisions, especially in Study 3, seldom reached the point at which the decision-makers' capacities for deliberate decision-making were actually overwhelmed. If capacities for deliberation were overwhelmed, decision-makers would have had no choice but to use a strongly intuitive mode. Comparing Studies 3 and 4, we found that for the most complex task environment in our study setup, a mass evacuation in Study 4, the reversal back to more intuition is more prominent overall. In addition, intuition is used earlier in the subjective complexity spectrum. In the highly complex mass evacuation training exercise, decision-makers more clearly demonstrated a strong return to intuitive decision-making once situational complexity reached a specific magnitude than in the comparatively less complex scenarios used in Study 3.

Although specific aspects, such as the location of the turning point or the degree to which intuition is relevant at the highest levels of complexity, vary somewhat between the studies, the general pattern of an inverse U-shaped relationship is remarkably stable. These findings suggest that previous studies on decision-making in complex task environments, which found a strong importance of intuitive or semi-intuitive (recognition-primed) decision-making (e.g., Zsambok & Klein, 1997), might have substantially overestimated the range of decisions for which intuitive decision-making is actually applied. In addition, the importance of the various degrees of complexity for the choice of decision modes might have been overlooked. The study that grounded the belief in a strong role of intuitive decision-making in "naturalistic" settings, an in-depth case study on decision-making of a firefighter during a challenging assignment (Klein et al., 2010), explicitly studied decisions under extreme time pressure. In such situations, deliberation is not an option to begin with as the deliberate decision-making process is too slow (Evans, 2008). Therefore, for decisions under extreme time pressure, it is not really a question of whether a decision-maker would use a deliberate or intuitive decision mode (Kahneman & Frederick, 2002).

However, this central contingency condition of the "naturalistic" decision-making stream got lost to some degree during the evolution of naturalistic decision-making research. Our results indicate that it might be time to remember this boundary condition, acknowledging that naturalistic decision-making in its focus on strongly intuitive decision situations empirically examines a fringe phenomenon. This stream of research helps us understand how proficient decision-makers remain actionable in highly challenging situations. Yet, it is not so well suited for deriving conclusions on how decision-makers approach decisions over varying degrees of complexity. Based on the empirical insights obtained in different studies, capturing decision mode and complexity with different measures (see Section 4.2), there is strong reason to doubt the general rule that proficient decision-makers mostly apply intuitive decision-making.

We actually found some evidence that contradicts the predictions of naturalistic decision-making. In particular, in Study 3 the highly experienced decision-makers were those that used intuition least (see Figure 4.5). However, this pattern looked slightly different in Study 4.

It is also important to note that although time pressure was an issue in most decisions, it very seldom reached extreme levels. Specifically, time pressure does not predict the choice of decision mode in our study, nor does it affect the other relationships that we examined. Typically, decision-makers have seconds or minutes to think about a decision in maritime SAR tasks, which is sufficient time to apply some degree of deliberate reasoning. We argue that this is also the case for the vast majority of judgmental decisions made in other complex task environments. Decisions under extreme time pressure do occur; and they certainly tend to be important, as the study of Klein et al. (2010) convincingly shows. However, these decisions are fairly specific and rare occurrences in the wide range of possible judgmental decisions in complex task environments. For those decisions that do not occur under extreme time pressure, our findings suggest that intuitive decision-making is only the standard approach for low-complexity decisions, which can be made based on using mature routine intuition.

Therefore, considering this relationship between situational complexity and the choice of decision mode in complex task environments, it seems likely that the choice of an appropriate level of deliberation follows a sufficing strategy. A deliberate decision mode is costly in terms of cognitive effort and opportunity costs (Kurzban et al., 2013). Therefore, decision-makers deliberate only as much as necessary in a given situation based on their perception of the decision environment. Situational complexity is thus a trigger of deliberation—that is,

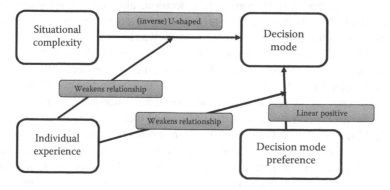

FIGURE 4.5 Observed effects of situational complexity, individual experience, and decision mode preference on decision mode.

increases in subjective environmental complexity trigger a more effortful deliberate decision mode. However, the complexity of a situation eventually exceeds decision-makers' capabilities for deliberation, resulting in a return to intuition under high situational complexity. This finding provides an important cornerstone for the identification of "use factors" of intuitive and deliberate decision modes (Dane & Pratt, 2007).

The second central insight refers to the important role of experience. In particular, we find that experience does not directly determine the decision mode in which decisions are made. However, it substantially and negatively affects the willingness or ability of decision-makers to adjust their decision mode to situational demands. In particular, experienced decision-makers do not display the (inverse) U-shaped relationship between situational complexity and decision mode discussed earlier. Instead, the relationship is linear and, in one study, even statistically insignificant for highly experienced decision-makers. This means that experienced decision-makers do not adjust their decision mode strongly based on subjective situational complexity. Instead, they tend to adhere to one particular decision mode irrespective of situational conditions. This finding refers to the idea that experienced decision-makers might be "entrenched" (Dane, 2010) in a particular decision mode, an insight that aligns with previous research indicating that experience comes with a loss of mental flexibility (Bilalić et al., 2010; Hecht & Proffitt, 1995; Lewandowsky & Kirsner, 2000).

Although it is beyond the abilities of our statistical approach to fully test the reasons and effects of this substantially reduced adjustment of decision mode to situational complexity, there is evidence that this effect might be negative for decision performance. We found that the negative relationship between complexity and satisfaction with a decision is significantly more pronounced for highly experienced as compared to less experienced decision-makers. Highly experienced decision-makers appear to experience a much stronger drop in terms of their ability to reach good decisions when complexity increases compared to less experienced decision-makers. This indicates that cognitive entrenchment might be a problem that actually decreases decision quality, although more research is required to strongly establish this relationship.

We found surprisingly little evidence that other individual-level variables play an important role. Stress resistance, in particular, appears to have no effect on choice of decision mode and satisfaction with a decision, although highly stress-resistant decision-makers tend to have a stronger tendency to prefer intuition. The organizational background (military vs. nonmilitary) as well as biographical variables also have a negligible effect. Mindfulness is somewhat more important, although the findings are not fully consistent. Specifically, mindfulness seems to help decision-makers adjust to situational demands. We found that decision-makers with low mindfulness as well as low experience followed their decision mode preference most strongly, indicating that these decision-makers find it difficult to adjust their decision mode to situational demands. Mindfulness is a focus on the current moment (Dane, 2011) and should thus help employees recognize and process cues. Based on this mechanism, we found that highly mindful decision-makers find it easier to successfully deviate from their decision mode preference. Still, as mentioned, these effects are clearly less consistent than the effects of situational complexity and experience.

In summary, the four survey studies we conducted in the field of maritime SAR substantially advanced our understanding of when and how decision-makers change their decision mode in complex task environments. In the following chapter, we turn our focus to our qualitative studies, which examine the question of how decisions are made in complex task environments based on in-depth case study methodology.

REFERENCES

Alter, A. L., Oppenheimer, D. M., Epley, N., & Eyre, R. N. 2007. Overcoming intuition: Metacognitive difficulty activates analytic reasoning. *Journal of Experimental Psychology. General*, 136(4): 569–576.

Bandura, A. 1977. Self-efficacy: Toward a unifying theory of behavioral change. *Psychological Review*, 84(2): 191–215.

Baylor, A. L. 2001. A U-shaped model for the development of intuition by level of expertise. *New Ideas in Psychology*, 19(3): 237–244.

Betsch, C. 2008. Chronic preferences for intuition and deliberation in decision making: Lessons learned about intuition from an individual differences approach. In H. Plessner, C. Betsch, & T. Betsch (Eds.), *Intuition in Judgement and Decision Making* (pp. 231–248). New York, London: Lawrence Erlbaum Associates Inc.

Betsch, T., & Glöckner, A. 2010. Intuition in judgment and decision making: Extensive thinking without effort. *Psychological Inquiry*, 21(4): 279–294.

Bilalić, M., McLeod, P., & Gobet, F. 2010. The mechanism of the *einstellung* (set) effect: A pervasive source of cognitive bias. *Current Directions in Psychological Science*, 19(2): 111–115.

Brown, K. W., & Ryan, R. M. 2003. The benefits of being present: Mindfulness and its role in psychological well-being. *Journal of Personality and Social Psychology*, 84(4): 822–848.

Campbell-Sills, L., & Stein, M. B. 2007. Psychometric analysis and refinement of the Connor–Davidson resilience scale (CD-RISC): Validation of a 10-item measure of resilience. *Journal of Traumatic Stress*, 20(6): 1019–1028.

Chreim, S., Williams, B., & Hinings, C. 2007. Interlevel influences on the reconstruction of professional role identity. *Academy of Management Journal*, 50(6): 1515–1539.

Dane, E. 2010. Reconsidering the trade-off between expertise and flexibility: A cognitive entrenchment perspective. *Academy of Management Review*, 35(4): 579–603.

Dane, E. 2011. Paying attention to mindfulness and its effects on task performance in the workplace. *Journal of Management*, 37(4): 991–1018.

Dane, E. 2013. Things seen and unseen: Investigating experience-based qualities of attention in a dynamic work setting. *Organization Studies*, 34(1): 45–78.

Dane, E., & Pratt, M. G. 2007. Exploring intuition and its role in managerial decision making. *Academy of Management Review*, 32(1): 33–54.

Dijksterhuis, A. 2004. Think different: The merits of unconscious thought in preference development and decision making. *Journal of Personality and Social Psychology*, 87(5): 586–598.

Dijkstra, K. A., van der Plight, J., & van Kleef, G. A. 2013. Deliberation vs. intuition: Decomposing the role of expertise in judgement and decision making. *Journal of Behavioral Decision Making*, 26(3): 285–294.

Epstein, S., Pacini, R., Denes-Raj, V., & Heier, H. 1996. Individual differences in intuitive-experiential and analytical-rational thinking styles. *Journal of Personality and Social Psychology*, 71(2): 390–405.

Evans, J. S. B. T. 2006. The heuristic-analytic theory of reasoning: Extensions and evalua-
    tion. *Psychological Bulletin & Review*, 13(3): 378–395.
Evans, J. S. B. T. 2008. Dual-processing accounts of reasoning, judgment, and social cogni-
    tion. *Annual Review of Psychology*, 59(1): 255–278.
Evans, J. S. B. T. 2010. Intuition and reasoning: A dual-process perspective. *Psychological
    Inquiry*, 21(4): 313–326.
Evans, J. S. B. T., & Stanovich, K. E. 2013. Dual-process theories of higher cognition:
    Advancing the debate. *Perspectives on Psychological Science*, 8(3): 223–241.
Gioia, D. A., & Poole, P. P. 1984. Scripts in organizational behavior. *Academy of Management
    Review*, 9(3): 449–459.
Goll, I., & Rasheed, A. A. 2005. The relationship between top management demographic
    characteristics, rational decision making, environmental munificence, and firm perfor-
    mance. *Organization Studies*, 26(7): 999–1023.
Hecht, H., & Proffitt, D. R. 1995. The price of expertise: Effects of experience on the water-
    level task. *Psychological Science*, 6(2): 90–95.
Kahneman, D., & Frederick, S. 2002. Representativeness revisited: Attribute substitution in
    intuitive judgment. In T. Gilovich, D. Griffin, & D. Kahneman (Eds.), *Heuristics and
    biases: The psychology of intuitive judgment* (pp. 49–81). Cambridge University Press.
Kahneman, D., & Klein, G. 2009. Conditions for intuitive expertise: A failure to disagree.
    American Psychologist, 64(6): 515–526.
Klein, G., Calderwood, R., & Clinton-Cirocco, A. 2010. Rapid decision making on the fire
    ground: The original study plus a postscript. *Journal of Cognitive Engineering and
    Decision Making*, 4(3): 186–209.
Klein, G. A. 1995. A recognition-primed decision (RPD) model of rapid decision making.
    In G. A. Klein, J. Orasanu, R. Calderwood, & C. E. Zsambok (Eds.), *Decision making
    in action. Models and methods* (2nd ed.) (pp. 138–148). Norwood: Ablex Publishing.
Kozlowski, S. W. J., & Ilgen, D. R. 2006. Enhancing the effectiveness of work groups and
    teams. *Psychological Science in the Public Interest*, 7(3): 77–124.
Kurzban, R., Duckworth, A., Kable, J. W., & Myers, J. 2013. An opportunity cost model
    of subjective effort and task performance. *The Behavioral and Brain Sciences*, 36(6):
    661–679.
Laureiro-Martinez, D. 2014. Cognitive control capabilities, routinization propensity, and
    decision-making performance. *Organization Science*, 25(4): 1111–1133.
Lazarus, R. 1993. From psychological stress to emotions: A history of changing outlooks.
    Annual Review of Psychology, 44: 1–21.
Lewandowsky, S., & Kirsner, K. 2000. Knowledge partitioning: Context-dependent use of
    expertise. *Memory & Cognition*, 28(2): 295–305.
Lieberman, M. D. 2000. Intuition: A social cognitive neuroscience approach. *Psychological
    Bulletin*, 126(1): 109–137.
Nemeth, C., & Klein, G. 2011. The naturalistic decision making perspective. *Wiley
    Encyclopedia of Operations Research and Management Science*.
Norrington, L., Quigley, J., Russell, A., & van der Meer, R. 2008. Modelling the reliability of
    search and rescue operations with Bayesian belief networks. *Reliability Engineering &
    System Safety*, 93(7): 940–949.
Plessner, H., Betsch, C., & Betsch, T. (Eds.) 2008. *Intuition in judgement and decision mak-
    ing*. New York, London: Lawrence Erlbaum Associates Inc.
Podsakoff, P. M. 1986. Self-reports in organizational research: Problems and prospects.
    Journal of Management, 12(4): 531–544.
Pratt, M. G., Rockmann, K. W., & Kaufmann, J. B. 2006. Constructing professional identity:
    The role of work and identity learning cycles in the customization of identity among
    medical residents. *Academy of Management Journal*, 49(2): 235–262.

Pretz, J. E. 2008. Intuition versus analysis: Strategy and experience in complex everyday problem solving. *Memory & Cognition*, 36(3): 554–566.

Scott, S. G., & Bruce, R. A. 1995. Decision-making style: The development and assessment of a new measure. *Educational and Psychological Measurement*, 55(5): 818–831.

Shiloh, S., Koren, S., & Zakay, D. 2001. Individual differences in compensatory decision-making style and need for closure as correlates of subjective decision complexity and difficulty. *Personality and Individual Differences*, 30(4): 699–710.

Steigenberger, N., Fuchs, H., & Lübcke, T. 2015. *Intuition or deliberation—How do professionals make decisions in action?* Academy of Management Proceedings, 11054.

Thompson, V. A., Prowse Turner, J. A., & Pennycook, G. 2011. Intuition, reason, and metacognition. *Cognitive Psychology*, 63(3): 107–140.

Zsambok, C. E., & Klein, G. A. 1997. *Naturalistic decision making*. Mahwah, NJ: Erlbaum.

5 Working in a Complex Maritime Task Environment
Case Studies

This chapter provides a practical, close-up view of the maritime field of decision-making. Two case studies provide quintessential insight into decision-making with respect to the organizational structures and several influencing factors. We chose two examples with different foci. The first case study (Section 5.1) is based on shadowing decision-makers in a large-scale, live SAR exercise, involving numerous SAR units and information sources. The second study (Section 5.2) consists of participant observations in SAR missions in the Aegean Sea, where an interorganizational crew rescued people from overcrowded, unseaworthy rubber boats. Whereas the first case study helps us understand decision-making processes more clearly, the second case study shows how these processes can develop in the first place. These case studies represent two different types of mass rescue operations. They provide good examples of the challenges that decision-makers face in complex maritime task environments.

5.1 A LOCAL INCIDENT COORDINATOR AT WORK: CASE STUDY ON DECISION-MAKING IN LARGE-SCALE, LIVE SAR EXERCISES

Live exercises are the most realistic but also the most expensive means to provide training in maritime SAR. In contrast to other training forms—for example, tabletop exercises and training in a simulator—an exercise on the open sea is neither meant to train only a specific aspect of search and rescue nor to be interrupted and repeated immediately. Instead, the interaction of different actors and participating SAR units forms the general purpose of such events. All aspects of a complex SAR assignment are conducted, observed, and debriefed to offer participants the opportunity to obtain lessons learned for each participant. Regardless of their level in the command structure, decision-makers are subject to all the effects they would face in real-life: weather, time pressure, mechanical and cognitive tasks using real tools, and so on. Such training is therefore as close to SAR reality as reasonably possible. We utilized two large-scale, live search and rescue exercises with nearly identical setups to follow decision-makers in a key position fulfilling their tasks.

5.1.1 INTRODUCTION

Shadowing a decision-maker in a large, live scenario is a seldom-experienced opportunity for researchers as well as SAR professionals. Due to the enormous effort necessary to plan and conduct a real-life exercise, these trainings are very rare events. If they also have an international component and include SAR services from different nations, the opportunities for live exercises become increasingly unlikely. Decision-making is a matter in SAR that can be observed on all levels in the command structure. Starting with hands-on decisions, such as if it would be better to wear a survival suit even if it is not strictly necessary, to tool-based decisions, such as by a doctor to prioritize the transportation of several injured persons over local treatment, or the creation of a communication plan by an on-scene coordinator (OSC)—each decision that is made during a SAR assignment may affect, to varying degrees, the success of the entire mission. Thus, there are always small and large decisions to make with different consequences to the mission as a whole. In Chapter 4, we provided an idea of how situational characteristics and personality characteristics affect decision-making. However, in order to obtain a deeper understanding of how these relationships play out in real life, a closer look at the key decision-makers is warranted. Obtaining such a closer look was the purpose of these case studies.

According to the International Maritime Organization, International Civil Aviation Organization (IAMSAR, 2010a,b, 2013), the main decision-makers in a maritime incident are the SAR mission coordinator (SMC) in the responsible rescue coordination center, the captain of the distressed vessel, and the OSC, who conducts all surface SAR activities, as well as the aircraft coordinator (ACO), who coordinates the air units on and over sea. The core of communications takes place between these actors: the SMC for the authorities ashore, the OSC for the surface SAR units and ships, the ACO for the SAR aircrafts, and the casualty vessel for their own distress situation. It is often the mission coordinator, the OSC, and the ACO who make the most influential decisions in maritime SAR missions, whereby the only decision-maker inside the distressed vessel is the captain of that vessel.

There is an ongoing debate on whether this setup is ideal; and different SAR services have begun to experiment with a function that is positioned between the captain of the distressed vessel and the OSC: the local incident coordinator (LICO). This function was created and named by instructors from the Naval Warfare School in Frederikshavn, Denmark (Møller, 2014)—introduced and tested in the Baltic SAREX 2014. The idea to add a subsidiary function to the international SAR structures stems from past incidents—at different places in the world—during which the captain or parts of the bridge crew abandoned ship, leaving no one on the bridge of the distressed vessel to coordinate the necessary measures. A distress situation is always unique and represents an extraordinary circumstance for a vessel and its crew; no one can predict how the personnel will behave and whether they are able to fulfill their jobs at any time. Thus, the idea came up to create a function that can offset the limitations that may occur on a distressed vessel's bridge.

5.1.1.1 The LICO Concept

A distress situation is an exceptional occurrence for a ship's crew and its master; safety drills and emergency plans were made to prepare and train the crew to practice, for example, damage control, medical first aid, and crowd management. In times in which relatively small crews handle large ships, the impact of an incident may become more serious with time due to the limited capacity of personnel to fight the damage. If parts of the crew are hit hard by the accident, the situation on board can worsen rapidly. The more diverse the required countermeasures are, the more external assistance is needed. If no external assistance is possible, the capabilities of a crew are likely to erode quickly and ineffectively.

A precondition to have a LICO in a rescue mission is the presence of another ship acting as an OSC for the distressed vessel. The SMC ashore will appoint an appropriate unit for the OSC tasks: a large SAR unit, authority vessel, or a military unit with adequate capabilities. In order to obtain an overview of the situation on board the vessel in distress, the OSC can appoint a LICO team and send it on board the casualty vessel. The head of this team should be the second in command (first mate) or another experienced person that the OSC trusts. Depending on the specific incident, additional qualifications may be useful to conduct the measures on board, for example, knowledge of hazardous materials, damage control, or firefighting. However, the most important task of a LICO is to act as a liaison and support the captain of the distressed vessel. To do so, he or she should use available knowledge on incident response management, the national SAR capabilities, national regulations, and IAMSAR guidelines. A LICO verifies the evaluation of the situation on board and makes sure that the available resources and assistance arriving can be used efficiently.

A suitable workplace for the LICO should be the bridge—if it is undamaged and still usable—near the captain and his* bridge crew. Depending on the situation on the bridge, the mentality of the captain, and the incident as a whole, the role of the LICO might be adaptable: Support can range from providing hints to the captain to more or less taking command. The exact form of support offered is the result of a negotiation process. It does not mean that the LICO takes over the responsibility for the ship. The captain always retains responsibility for his crew and ship. For example, it is still the captain's responsibility to decide when a ship should be abandoned in the worst circumstances. The LICO's responsibility is the safety of the rescue teams on board the casualty vessel. If he makes the decision that it is necessary to leave the ship, and the captain does not agree, the LICO can give the order to all rescue teams on board to abandon the ship. Thus, his role is more than just that of a supporter: A LICO is a consultant with his own staff, consisting of crew members from the OSC's ship as well as other units such as medical teams or firefighters.

In practice, an intensive exchange of ideas and information between the captain and the LICO takes place; they talk about all the necessary measures to take on board as well as how and from where resources can come. At the same time, different knowledge and experiences come together to develop a plan based on the information available at that time. Periodical situation reports (sitreps) from the LICO to the mother ship and specific requests for assistance ensure that the OSC always has an overview of the situation on board. During various real-life SAR exercises in Denmark, the creators of the LICO concept received positive feedback from the

---

* For reasons of brevity, we use the masculine form for the different decision-makers, including also female captains, ship masters, LICOs, and other personnel.

ships' masters they trained with. Having the support of a highly skilled person during a distress situation was perceived as absolutely positive and was highly valued.

A LICO is an interesting decision-maker in maritime SAR: The LICO is a SAR professional who is physically and mentally fully involved in the distress situation. All activities initiated by this person take place on the vessel in distress; and he often faces the maximum complexity the mission has to offer. The LICO team, itself, is in the place of danger with the responsibility for all rescue forces on board the casualty vessel. Therefore, a look over the LICO's shoulder—a rare view, also for the involved rescue personnel—provides us with a profound picture of what a decision-maker faces in highly complex tasks. We can understand how the decision-maker approaches decisions, depending on environmental factors, including an in-depth understanding of how the specific situation, developed from the situation and decisions made earlier by the decision-maker or others, influences decision-making. Thus, we follow how the LICO tasks are handled by decision-makers with different levels of experience.

5.1.2 THE EXERCISE

The mass rescue operation scenario was the main event at the Baltic SAREX and took place on the fourth day of the exercise week. All the trainings and exercise days prior act as preparation for the variety of tasks and measures that must be made during a mass rescue at sea. We studied the Baltic SAREX 2014 and 2015.

5.1.2.1 General Scenario

To create a challenging scenario for all of the more than 20 participating SAR units at the Baltic SAREX, a collisions scenario was created that began with a failed VHF communication between two ships and immediately following mayday calls on the VHF radio.* A ferry with numerous persons on board collided with a freight ship and threatened to capsize. It took on water and had several injured persons on board as well as persons in the water. The rescue coordination center answered the mayday call, asked for the position of the ship in distress, and appointed the OSC for this incident—the person who is responsible for the SAR measures for both collided ships. The mission began with little information about the situation of the collided vessels. The other ships in the vicinity—here, the participating SAR units—called in, according to international regulations, and offered their support. From the moment of being appointed, the OSC took over responsibility for coordinating all assistance measures at the incident. For the Baltic SAREX mass rescue operation, this meant "free game," in that the exercise staff provided a maximum of independence to the participants during the scenario. The only direct link to the exercise was by the rescue center that was manned with acting SMCs, who were able to add information if necessary and influence the activities on-scene.

---

* A codebook that was part of the exercise order ensured that the official distress channel 16 was not used; and other alerting keywords were changed in the exercise communication. For example, the word "life raft" was changed to "bike" and "person" became "soul" to prevent ships not involved in the exercise from accidentally overhearing the radio communication and believing that a real emergency was taking place.

FIGURE 5.1   The Bridge Windows on the OSC Vessel During Baltic SAREX 2014.

Periodically sent sitreps from the OSC to the SMC reported the relevant outcomes and the situation on-scene. A standard sitrep template from the IAMSAR Vol. III, Appendix Delta (IAMSAR, 2010a) can be used for this purpose. This manual is also the basic guideline that can be utilized by all participants as a reference book during an exercise. Additional technical equipment and gear can be used as available, for example, forward-looking infrared for searching for persons in the water, (electronic) sea charts for maneuvering, or minesweeping software to monitor the coverage of a search area. Often, whiteboards and flip charts are utilized to visualize the situation in command and control tasks to establish the head count and retain an overview of available resources. Over the years, the bridge windows have become more and more important for documentation as a space to write new facts with whiteboard markers (Figure 5.1).

5.1.2.2   Preliminary Workshops

Academic workshops are an important part of the Baltic SAREX, at which useful tools and lessons learned for the commanding officers and seconds in command are offered. In 2014 and 2015, two workshops related more or less directly to decision-making. The first workshop was on a tool that supports structured decision-making, providing a structure with which decision-makers can separate facts from assumptions and threats and measures from orders. This tool came from firefighting on land and was transferred to maritime incidents. It is based on established military strategy and tactics from Carl von Clausewitz and Helmuth von Moltke such as mission command. The tool aims to help the users prepare all gathered information on the incident in a structured way. The tool itself was not uncontroversial among the SAR experts; some said that the necessary usability was not given and others that the tool is more suitable for disasters on a national level rather than most maritime accidents. For the purpose of our study, the method is worth mentioning for two reasons: (1) Offering such a workshop is an indication that there is both a lack of tools and a practical need for decision support in the maritime SAR field and (2) a tool like this—supporting systematic information processing—encourages deliberate decision-making, which appears to be welcome and accepted in the maritime SAR domain. We will discuss the second point, in particular, more extensively in Chapter 6.

Another workshop held during Baltic SAREX 2015 was a LICO workshop to prepare participants for related tasks in a mass rescue operation. In this workshop, lessons learned from 2014 were also discussed. This is a distinguishing characteristic of high-reliability contexts: Learning opportunities are used to subsequently improve aspects such as procedures, equipment, and communications. In this way, previous learning by individuals is made available to a broader set of potential users.

5.1.2.3 Methods

To capture most decisions that the LICO makes during the assignment, we conducted two shadowing studies that allowed us to derive in-depth insights into decision-making in complex task environments (McDonald, 2005). Both shadowing studies referred to mass rescue missions at the Baltic SAREX: the first was conducted in 2014, the second in 2015. For our first shadowing study, we had a preparation period of two years, which offered us the opportunity to carefully prepare and test the shadowing methods regarding the research process, technical equipment, understanding of the SAR organization that was built ad hoc by the OSC, and artifacts that were utilized by the LICO. In the end, we developed a multivideo approach, which we combined with expert interviews to fully utilize the information we obtained in the videos. Specifically, we watched the videos with field experts to obtain their interpretation of occurrences during the mission.

We used outdoor action cameras to capture the LICO activities as well as actions in other parts of the distressed vessel. To obtain a subjective view of the situation, the LICO wore a chest harness with an action camera. Prior to the study, we conducted several trials to test camera settings, positioning, and technical maintenance of the research equipment and decided the chest harness setup was the most appropriate for our purpose to conduct video-based research in dynamic environments (see Lübcke, 2016). Additional fixed cameras near the expected workplace on the target ship filmed the activities from an observer position. Furthermore, an additional person on board wore a harness camera to deliver selective insights from the hot spots: gathering places, hoist points, and firefighting activities. In the end, we had four simultaneous recordings from 2014 and six simultaneous recordings from 2015.

The recordings were edited by a professional cutter to form a synchronized nonstop analysis video with indications of local time and time after the distress call. This analysis video was the basis for the expert discussion and evaluation afterward. Figures 5.2 and 5.3 are exemplary screenshots from the videos.

For video interpretation, we used a semistandardized questionnaire to allow experts to (1) identify the LICO's decisions as they perceived them and (2) evaluate decision quality and the decision environment, for example, in terms of time

FIGURE 5.2 Screenshot from Analysis Video of 2014 Study.

pressure, information (overload), distraction, and predictability. These expert sessions lasted between three and four hours, in which 11 decisions from one decision-maker were evaluated. The interviews were documented on a paper questionnaire and additionally audiorecorded to capture all annotations that were not covered in the questionnaire. We conducted all our expert interviews in English. The expert interviews then guided our own interpretations of the video data.

Each of our experts had more than 20 years of practical experience in their maritime SAR career and have sailed in various positions on board rescue units, worked as instructors to train SMCs and OSCs, developed national and international guidelines for SAR matters, worked as investigators on several very serious maritime casualties, and worked as observers and umpires in real-life exercises. All experts volunteered their time for the expert interviews. As compensation, they received confidential copies of our analysis videos for training purposes.

5.1.3 OBSERVATIONS

Due to the free-game concept of the mass rescue scenario, we began our observation of the LICO when he embarked the distressed vessel. One of the researchers followed the VHF radio communication and, as soon as the estimated time of the LICO team's arrival was announced, he moved to the pilot door in expectation of the first contact. The harness camera was activated and prepared for the LICO. One of the authors followed the LICO afterward to make notes and take photographs from a background position as well as maintain all cameras. We finished the video recordings and observations when the exercise was declared executed. This setup was used for both the 2014 and 2015 study.

FIGURE 5.3   Screenshot from Analysis Video of 2015 Study.

5.1.3.1   LICO Shadowing Study 2014 (Team 1)*

The mass rescue exercise in 2014 took place under quite calm weather conditions: mostly cloudy, in the beginning scattered showers, but later sunny, light wind, and good visibility. At the beginning of our case analysis, we will look at the first half hour of the LICO's presence on board the vessel in distress.

Approximately 39 minutes after the distress call, a rubber boat with the LICO arrived at the distressed ferry. They boarded the vessel via the pilot door to the car deck, which was held open with a launched Jacob's ladder but not manned at that time. On the car deck, crew members on the ferry were involved in firefighting measures somewhere in a staircase behind an open fire door, 60 meters away from the pilot door. Several persons were sitting or lying coughing on the car deck's floor; one person was doing CPR. Three persons from the rubber boat entered the ship through the pilot door: the person designated to be the LICO; an assistant to the LICO; and a third person, the so-called tallyman, whose job was to make the head count. They walked together to the group of crew members from the ferry near the staircase and asked them in Danish how to get to the bridge. The chief engineer of the ferry told them—also in Danish—that they have serious problems with firefighting capacities and several persons with smoke inhalation. The LICO entrusted his assistant to stay at the car deck to have him for liaison—both the LICO and assistant had military VHF radios—and to conduct the firefighting in collaboration with the ferry's chief engineer. After that, he continued to the bridge guided by one of the engineers (and followed by observers), using a hidden staircase near the bow (at the opposite side) of the ship. Underway, he gave a brief sitrep to his commanding officer on the mother ship. Two minutes later, he arrived at the bridge, introduced himself in Danish to the captain as the LICO, and received a brief verbal report of the general situation (see Figure 5.4).

Subsequently, he asked the captain about the location of the gathering point for injured persons and requested a fire and safety plan for the ship. The captain began to guide him to a rolled-out safety plan on a wheelhouse wall. However, the LICO first approached one of the bridge windows and placed his writing case and radios on the windowsill (see Figure 5.5). Then, he heard a VHF call from his mother ship, took the VHF radio device, answered, and then walked to the safety plan—while continuing to talk on his radio—where the captain was waiting.

The captain explained the vessel to him as well as where the gathering points were located. After a brief conversation on the number of persons on board the ferry, the LICO called the OSC via VHF radio and forwarded this number to him and then asked for firefighting assistance. Subsequently, he used his military VHF radio for further communications and gave instructions regarding the boarding of firefighters. He did all this wearing his survival suit and handling two VHF radios: the military one in the left hand and the civilian one in the right hand. Thereafter, he instructed the ferry's first officer to contact the crew members on the car deck about the forthcoming arrival of firefighters.

---

* For reasons of brevity, the LICO team we observed in 2014 will be named Team 1 and the team from 2015 Team 2.

FIGURE 5.4 Introduction to the Situation Aboard the Vessel in Distress (Team 1).

FIGURE 5.5 The LICO Working Space on the Windowsill (Team 1).

Three minutes later, after a brief conversation with the captain about the parallel ongoing hoisting from the stern of the ship, the LICO asked his commanding officer via military VHF how many people would be delivered from the OSC's vessel. Meanwhile, the first and second officers were supporting him by commanding the hoist operation, assisting in the head count, and communicating with the other decks of the ferry. All gathered information was noted by the LICO in his writing case on the windowsill. There were three additional decisions that the LICO made afterward. He instructed the persons on the car deck to ensure the head count at the pilot door. A few minutes later, he asked the OSC to allot a free VHF channel to him. Finally, approximately 25 minutes after entering the casualty vessel, he informed his commanding officer on the Danish vessel about the number of persons on board the ferry.

The LICO worked almost four hours on the bridge of the vessel in distress and made hundreds of decisions at different points in time; we will focus our discussion on the decisions made within the LICO's first 25 minutes on board the ferry.

There are two main factors to point out here that made the LICO's work much more difficult than it had to be and were not under his influence. First, the experienced OSC from Finland decided at a very early stage to delegate responsibilities for the measures on a Danish ferry in distress to a Danish rescue vessel as sub-OSC and issued them the order to send a LICO. From the human factors perspective, this was a smart decision to avoid misunderstandings between the rescue units and the vessel in distress. Yet, this decision contradicted in the end the concept of the LICO, who should be an OSC's person of trust. The direct link between the OSC and the LICO enables direct access to all available resources without communication via third parties. However, in the case of this mass evacuation, such a direct link was not in place. Instead, during the whole scenario, the communication structure between these two key actors was ambiguous, in particular as the LICO had two superiors to report to: the OSC and his own captain. The OSC failed to make sure that the Danish LICO reported and communicated directly to him. The LICO, in turn, was so focused on including his commanding officer that a large amount of redundant communication made the structure unnecessarily busy. Furthermore, this communication increased uncertainty more than necessary, as no one knew what he or she could do or how extensive his or her own competencies were.

A related factor was the preexisting hierarchy, which affected the collaboration between the LICO and his mother ship, who was the sub-OSC for the distressed ferry: The sub-OSC sent his second in command on board the ferry, following the typical LICO setup. This opened up an area of subtle conflict between the LICO—a young lieutenant—and his commanding officer on the mother ship. Being a coordinator, the LICO, the second in command, should switch to another role as should his commanding officer. In these new roles, the second in command should request assistance or even give orders to his commanding officer, who ideally would have to follow these orders to grant the best working conditions for his liaison officer on the distressed vessel. Unfortunately, examining the working conditions for the LICO, this switch from rank- to role-based hierarchy did not take place. Instead, it was a tightrope walk to be aware of the interdependences among both roles and positions as well as their consequences to the mission's success.*

These two factors should be kept in mind when we comparatively discuss the specific decisions later.

5.1.3.2 LICO Shadowing 2015 (Team 2)

In the 2015 mass rescue scenario, the LICO team arrived approximately 35 minutes after the alert and embarked the distressed ship via the pilot door that was manned

---

* This issue can be remedied by training, for example, by offering the persons in charge opportunities to practice servant leadership, which does not question their rank but still opens up new possibilities to recognize responsibility.

FIGURE 5.6   Meeting Role Players Acting as Casualties (Team 2).

with two crew members. The weather conditions were a little rougher than the year before: sea state of 1.5 meters on average,* windy, and scattered showers. The visibility conditions were moderate. Like the year before, coincidentally, the team also came from a Danish naval vessel. After boarding the vessel, the LICO asked one of the crew members in English for the captain. This person guided the LICO and his assistant toward the aft of the ship and then up several flights of stairs to the bridge. Leaving the boarding zone, the LICO informed the third person on his team—the tallyman—in Danish about the VHF channel on which they could be contacted. On the way to the stairs, while on the open deck, some role players acting as passengers attempted to catch the LICO's attention (see Figure 5.6). He told them in English that the team would come back and help them, but first they had to go to the captain. Then they quickly continued to the bridge.

Underway, the LICO communicated via VHF radio, while the assistant followed him with a backpack. Arriving on the bridge, he introduced the group as a "Danish rescue team," asked for the captain, and told him that they came to help him to deal with a fire they had on board. Then, both removed their safety equipment (helmets and survival suits). For the scenario, they wore pink sports tricots marked as "LICO" and "LICO Assistant." Under the semitranslucent tricot, the epaulettes were visible: The LICO was a master chief petty officer, whereas his assistant was a young lieutenant. After an umpire questioned—during a brief break—whether the hierarchy during the exercise should not be the reverse, the lieutenant answered with "Experience matters most!"

Back in the exercise, the LICO ensured that his assistant had the correct VHF channel to communicate with their teammate at the pilot door. Afterward, he turned to the captain and asked him for a sketch of the ship. He received a fire and safety plan. Then, the captain began explaining details on the accident. However, the LICO interrupted him with the words "Wait for my assistant, please," who was conducting

---

* The Baltic Sea has short and rough swell that particularly affects the smaller units and dinghies.

FIGURE 5.7   Introduction on the Bridge (Team 2).

a communications check in Danish at the same time. After he finished this communication, the assistant turned to the chart table, where the captain and the LICO were waiting for him. After ensuring that everyone was ready, the LICO asked the captain to explain what happened on board. During the explanations from the captain, the LICO instructed his assistant to take notes (see Figure 5.7).

As the captain seemed to offer too much irrelevant information, the LICO brought the talk back to the most pressing topics. The captain handed over a passenger list. At the same time, another person in a Danish Navy survival suit entered the bridge and delivered to the assistant an additional backpack filled mainly with walkie-talkies. This person was later given the task of a runner—that is, to be eyewitness and liaison officer to the different teams on board. All the activities explained above took place within less than seven minutes after boarding the distressed vessel.

5.1.3.3   Comparison

As previously mentioned, the scenarios of the mass rescue exercise were nearly the same in both years and observations. Due to the setup, the rescue team approached the distressed vessel quite quickly—in less than 40 minutes after the distress call. This time frame is certainly not representative of all real-life SAR missions. Rather, it is more a marker for us to have a stable framework in both inquiries. The team we observed in 2015 arrived at the ship four minutes earlier than the team the year before, which may have opened up opportunities, for example, related to the fact that the pilot door was still manned and that the team was not distracted by an ongoing firefighting operation nearby. Team 2 was earlier because of decisions made elsewhere: Above, we discussed the smart but tricky decision from the OSC in the 2014 exercise requesting a sub-OSC to provide a LICO team. That was different in 2015, when the LICO team was from the OSC's crew. In contrast to 2014, when a Finnish vessel was in charge as OSC, in 2015 the mandate was given to a nearby Danish

Navy vessel. In addition, the OSC was already closer to the casualty than the OSC in the 2014 exercise had been. Both variables had an influence on the reduced time required to deliver a boarding team to the vessel in distress in 2015.

In both years, we observed teams of Danish marine officers. They all came from picket boats of the same category with similar technological and human capabilities. The teams had roughly the same educational background as well as the same legal framework within which they work—for instance, the national SAR Denmark regulations—and were trained according the *Guide to Coordination of Major SAR Incidents at Sea* (Møller, 2014). Accordingly, the initial situation in both inquiries was nearly identical with the exception that the preliminary decisions from the currently responsible OSC created a unique command and control setting for each mission.

Both came to the same final decision to evacuate all passengers from the ship. However, workplace setup, organization, and the way toward this decision differed substantially. This shows that there are typically different paths toward the same conclusion. Multioptionality, particularly in the beginning of a mission, forces the decision-maker to create a certain stage from which he conducts the next steps. In our two observations, it was apparent that the teams did this very differently. Let us first look at how the teams established a suitable command post (the bridge). Team 1 immediately began to make tactical decisions and worked in command-and-control mode nearly from the first moment. As Team 2, they wanted to go to the bridge immediately without knowing the ship. Team 2 had the advantage of finding a manned pilot door; thus, they found it easy to ask someone for directions to the bridge. Team 1 had no one waiting at the pilot door, so this option was not available. After boarding the ferry, they looked around and went to the next available person near the staircase at the aft, hoping to encounter a crew member to ask. Here, the decision was more complicated, as cues were sparse and outside information was not as readily available. The team went straightaway to the persons at the farther end of the car deck. In doing so, they did not consider an outside piece of information that would have been available to them, the metal holder labeled "fire safety plan" next to the door (see Figure 5.8). Perhaps the presence of observers waiting for their next actions increased stress and subjective time pressure to the team members. However, short consideration should have reminded the professionals that fire plans are mandatory on a passenger vessel (according to SOLAS, 1974, as amended) and could be available near places where firefighters would also board the ship.

In the next stage, the teams were directly confronted with the distress situation aboard, being "irritated" by ongoing operational events. Team 1 was faced with ongoing firefighting activities, injured persons, and resuscitation attempts on the car deck. Team 2 ran into disoriented passengers shaking with cold and asking for help. The LICO in Team 2 answered the persons politely but firmly that they would be helped but that the rescue team had to go to the bridge first. Contacting the captain was the highest priority based on the notion that immediate hands-on work is not part of the mission leader's role, which is focused on delegation, evaluation, and control.

Team 1 handled the first "irritation" differently. They approached the persons on the car deck and asked the chief engineer for directions to the bridge. Instead of answering this question, the chief engineer told the LICO about their limited firefighting

FIGURE 5.8 Exit to Pilot Door with Fire Safety Plan (Dark Colored Holder) near the Door (Team 1).

capabilities and the injured persons that they had to care for. The team leader decided to leave his assistant there and delegated him to command the firefighting. In the expert interviews, this was the only decision for which the experts held contrary opinions: Some held the view that this was a good decision, referring to the LICO's responsibility to delegate and give orders. He did that and solved two problems: He had a person of trust (his assistant) on the car deck and a reliable (military) VHF connection to the firefighting team at the same time. Others were convinced that this was a bad decision, arguing that this tactical decision had a negative impact on all that could come later. Not having an assistant later made it more difficult to fulfill his future tasks. He could have held the firefighting activities near the car deck in mind and arrange VHF devices for them later from the bridge or from his mother ship.

From a decision mode perspective, the decision was made with an approach combining deliberate and intuitive elements. In the decision situation, he was distracted by ongoing radio communication between the OSC and another vessel, receiving information about the situation aboard the ferry from a nearly overwhelmed chief engineer in Danish, and visual impacts from resuscitation attempts in the background. He consciously tried to collect all available information to decide deliberately. The distracting impressions made the decision-making process highly complex. He then seemed to eventually give up on deliberation and went with a partially intuitive decision. In the end, this resulted in an acceptable decision that might have consequences later, as it implied committing an important resource early on—that is, his assistant.

Thus, Team 1 consisted of only the LICO himself when arriving on the bridge, where he immediately began his investigations. Quite structured, he inquired about the necessary information to obtain an overview of the situation aboard. Doing this, he showed deliberate decision behavior, collecting and structuring outside information. However, he did not decide deliberately in all events, in particular at times when he got distracted. This was, in particular, the case when the VHF communication and documentation made him very busy (see Figure 5.9). Due to the uncertainty regarding with whom he must communicate—with the Finnish OSC, his mother ship, or both—and what his position in the command system was, much unnecessary communication took place that made his situation much more complex and misled decision-making at some points. Although he tried to decide as deliberately as possible, in demanding situations he used a more intuitive mode, clearly driven by the overwhelming situation.

Contrary to Team 1, the LICO of Team 2 arrived with his assistant on the bridge. Team 2 started preparing themselves first and then organized their workplace. They took off their survival suits, placed their VHF radios on the chart table as well as some prepared charts to track the available resources, damages, and heads. The assistant was prepared with a red and a black crayon and a pencil-style eraser in his hand to make the documentation. During nearly the entire mission, the duo of LICO and assistant was working at the chart table, where the fire safety plan was placed on the left and the documentation on the right side of the table (see Figure 5.10). Each team member had his own VHF radio for communication. The LICO conducted all external communications with the OSC, whereas the assistant held contact to the tallyman at the pilot door and another team member who acted as runner.

Arriving rescue teams from other SAR units had to come to the bridge first to receive their orders from the LICO face-to-face. Doing so, the LICO was able to understand the capabilities of the rescue teams and could adjust his plans if necessary.

FIGURE 5.9 Intensive Radio Communications (Team 1).

FIGURE 5.10 Workplace Arrangement (Team 2).

Furthermore, the new rescue team was informed about who the local mission leader was as well as that they were conducted by the assistant via the assigned VHF channel.

Due to this way of working, a misunderstanding in tactics was solved before it had a negative influence on the rescue measures. A firefighting force arrived on board the vessel in distress. The LICO received information that the firefighting team consisted of 10 persons who were also qualified paramedics. His plan was to divide the group into two teams: one for firefighting and the other to care for injured persons in another part of the ship. When the head of the firefighters arrived at the bridge, the LICO issued an order according to this idea. However, the head of the firefighting teams rejected the order. The LICO was afraid that the firefighter was questioning his authority until the firefighter explained that they had arrived with only five men as well as how the firefighting team works: their operating procedures require having the full team for each task, so they would be able to offer firefighting and medical assistance but not both at the same time. Thus, the LICO had all necessary facts he needed to make a good decision.

The modus operandi established by Team 2 continues the consistently practiced deliberate decision-making of this team. Most decisions were aligned between the LICO and his assistant; thus, both always had the same overview of the situation and the decisions being made. Team 2 also always spoke with one voice. When the assistant was called on VHF or made a call, he referred to himself with the Danish word *indsatsleder*, which means incident commander. If someone called on his radio for the *indsatsleder*, then the LICO approached his assistant: "You have a call!" It was a big advantage for Team 2 to be able to work in a team, creating for themselves a proper work environment and sharing workload and responsibilities. This helped them increase the cognitive capacities available to them and avoid bottleneck problems, which would otherwise have caused their deliberate decision making approach to become dysfunctional.

In contrast, the LICO of Team 1 gave up the opportunity to have an assistant on the bridge to share ideas and workload early in the mission. He created his workplace setup ad hoc: The writing map was placed on a windowsill at the starboard side about four meters away from the fire safety plan that was rolled down at the rear wall of the bridge (see Figure 5.11). In addition to the writing map, he had two walkie-talkies, a mobile phone, two ball pens, a fineliner, and a text marker.

As mentioned earlier, he immediately began his investigations to obtain all available information, which led to substantial cognitive workload. Compared with Team 2 that worked from the chart table all the time, the LICO walked around on the bridge. He often went to the windowsill with his writing map and the roll-down fire safety plan on one of the ledgers, but most of the time, he was running around with a radio in his hands. His work was even more difficult in that the captain was not particularly active in his role. The captain answered questions when someone asked him, but most of the time, he remained relatively passive in the background. In the absence of an assistant, the LICO slowly included the bridge crew in the workload. The first officer was already in contact with other crew members in all parts of the ferry—for instance, also with the chief engineer conducting the firefighting near the car deck. The LICO working next to the first officer was in contact with his assistant, who was conducting firefighting together with the chief engineer, by military VHF radio. This alleged redundancy made the communication situation aboard the ferry more complex than necessary.

The LICO involved the bridge crew step by step in command and control due to the lack of personnel resources, yet the involved parties did not talk about how they should organize themselves; therefore, the resulting ad hoc team emerged from a relatively unstructured social process. The bridge officers merely continued to discharge their tasks out of the roles they typically had on board. In terms of the topics and information shared by each bridge officer, the LICO obtained an idea of the

FIGURE 5.11   Workplace Arrangement (Team 1).

established division of work and adjusted his working to the conditions he found. He, himself, compensated for all the functional gaps he recognized in the ad hoc structure. He did so successfully, but this structure led to a high workload. He noted all relevant facts in his writing map, made sketches to visualize the situation for himself, and strove to conduct the rescue measures, so whenever possible he applied a deliberate decision mode.

The establishment of the ad hoc incident coordination team consisting of the LICO and the bridge officers occurred in an intuitive manner. Switching to a relatively intuitive mode, he saved time and adjusted himself to the given conditions, which was the response to a problem that could have been avoided in the first place: The entire time, they had at least two parallel communication chains aboard the vessel in distress—the ship's own communications structure conducted by the bridge officers and an additional structure via military radios between the LICO, his assistant on board the ferry, and the commanding officer on the mother ship. In addition, when additional rescue forces from other ships boarded the ferry and had their own communications gear with them, a substantial amount of information ran around in circles but did not necessarily arrive at the relevant person. As a result, the communication conditions in the decision-making environment were quite complex.

Table 5.1 outlines how decisions were approached by the two teams.

Overall, we must consider that the conditions under which Team 1 had to make decisions were more complex than those faced by Team 2. It is evident that the complexity in Team 1's mission was mainly created by the decision-maker himself as well as other decision-makers in the exercise: on the basis of an OSC decision to give the sub-OSC the responsibility to send a LICO team to the vessel in distress. During the mission, no one clarified the command structure, most notably regarding the well-trained but less experienced LICO. He tried to make the best of the

TABLE 5.1
Comparison of Decision Approaches between Team 1 and Team 2

Phase	Team 1	Decision Mode	Team 2	Decision Mode
Boarding	Searching for crew members	I/D	Asking crew members at the pilot door	D
Being confronted with casualties/situation on board	Gets involved in local situation and starts conducting direct measures	I/D	Focuses on the tasks/ mission and postpones direct measures	D
Activities after arrival at the bridge	Begins investigations	D	Prepares workplace	D
	Communicates with OSC and mother ship	I		
Workplace setup	Ad hoc, scurrying around on the bridge	I	Arrives with prepared, appropriate gear	D
Conducting rescue measures	Involves the bridge crew, ad hoc team	I/D	Structured collaboration with all stakeholders	D

situation but was not encouraged to enforce clarifying this problem or overruling somebody for the purpose of mission success; neither by his commanding officer on the mother ship nor the bridge officers on the ferry. In contrast, Team 2 decided in a predominantly deliberate manner at all times, which was made possible by reserving cognitive resources and avoiding any situation that would force the team toward a more intuitive mode.

5.1.4 Resumé and Complementary Observations

With Team 1, we followed a relatively inexperienced decision-maker who started his operations with a lack of resources and information. After an early decision, while attempting to find a quick tactical solution, he unwittingly created highly demanding conditions for his subsequent tasks. He managed to solve the decision problems solidly in the following, using a mixture of deliberation and intuition in his tactical decisions. He did not have the means to compensate for his early decision in a manner that did not wear himself out. However, he was continually able to adjust to changing conditions when necessary. The other observed team (Team 2) was better prepared, better equipped, had more team members, and had an experienced leader who was not the highest in rank but the most experienced available. This team demonstrated how to use the scope in an uncertain situation to create working conditions for consistent decision-making. Considering the general distribution of decision modes, we see here the picture already identified in Chapter 4; whereas the less experienced decision-maker adjusted his decision mode quite substantially to the situation, the highly experienced decision-maker did so only to a small extent and, instead, followed a specific approach. In particular, this case is an example of how strongly situational characteristics shape decision-making in complex task environments.

5.1.4.1 Complexity as a Framework for Decision-Making

In our observations, we understand that complexity in the maritime domain is not limited to poor weather conditions, for example, high sea state, waves, or poor visibility, and their effects on the seamen and rescue workers, such as fatigue and seasickness. Complexity is also increased by human capabilities—the resources out of the capabilities of their own crew as well as the capabilities of other involved parties—and the hardware of the units—rescue cruisers and other assisting vessels—and their behavior also influence complexity. If unknown units are involved, the uncertainty regarding their capabilities concerning seaworthiness and maneuvering may increase the complexity of a decision situation for a decision-maker in a coordinating role. Also, poor or, especially, false information or prior decisions affect complexity.

Experience plays a critical role in decision-making in complex tasks, and the following story, told in one of our expert interviews, is a good example of this:

> If you have (...) (...experienced and unexperienced) sailors on a small vessel (...), they will probably make decisions like "We can't go there, the weather is too bad!" And then some of the other guys would say "What weather?" But then again, you have these

old skippers with gray beards and a big pipe and all those typical things, and he'd say "Of course we can go." And then maybe half of this search and rescue unit will think, "The guy in charge of us is bloody mad. He says that we have to go over there." (...) So the experienced captain has to think about that. He has got people around him, helping him, that maybe have a different view of how the situation is. That goes for the whole group of rescue units—also for your ship. As a captain, when I was skipper on a boat, I was sometimes like "This is no problem...," but then I had cadets there shitting their pants or whatever they were doing. And they were like "Skipper, are we gonna die?" I'd be like "No! Except if you are really ill! Why are you asking me that question?" "This must be the worst weather you ever saw!" "We're not even wet yet." You have a different perspective of things. (...) I would not say everybody can do this. I also have guys with less experience, so I have to make a less hard decision or you don't sail the ship that hard. Don't demand the same from your people. So, experience as a sailor has a big effect and it goes both ways.

It is important to know the capabilities and limitations of personnel and hardware in certain situations as those determine the possibilities you have in your decision-making. Subjective complexity depends on the experience of the individual. Thus, experience also makes a difference in terms of the judgment of situations due to perceived complexity.

When asked about the impact of increasing complexity on decision-making, an exemplary expert's statement descriptively explained what happens:

I think it will change the style that is used. And I think maybe if the conditions are not so good, I think the on-scene coordinator will switch to intuitive decisions. Because he will be pressed to the wall and then you will not use the rational tool to make some decisions. Sometimes, (I wish) we could force him to use checklists.

This and comparable statements also reflect the organizational preference for deliberation, which we will discuss in the following.

5.1.4.2 Appropriate Decision Modes and Decision Quality

Overall, we saw that especially the deliberate decision mode appears to be approved in the maritime SAR domain. In both inquiries, we observed that the decision-makers strived to decide deliberately all the time. Decision support tools, such as the ones discussed in a workshop at the beginning of the exercise (see Section 5.1.2), are designed to facilitate this decision-making approach. Yet, they were used to only a limited extent. A few catchwords were consulted to have a general overview of the situation aboard, but the decision support tools were never fully implemented. Also, the experts pointed out in the interviews that they are convinced that deliberate decision-making is of the highest relevance in maritime SAR. At the same time, they stated that decision support tools, such as the one described earlier, are too cognitively demanding to be suitable for the complexity aboard a ship in a mass rescue operation. It may be useful on land, such as in a rescue planning team, but not in the field. That deliberate decision-making has its advantages is certain, in particular, to reproduce decision rules for later investigations or when it is necessary to hand the task over to another decision-maker. However, we have already seen that this is only one part of reality; intuitive decision-making is equally present. After asking experts about the appropriate decision-mode—deliberative or intuitive—all agreed:

> In real life, both (intuitive as well as deliberate decision making) is needed, depending on the situation you are in—how complex the situation is. (…) A lot of SAR missions are not very well known from the beginning. Many things change during the incident; many small details pop up. Reflecting on the situation—you have to be intuitive all the time. The situation changes all the time, and when you get information, you might be in the situation that you have to change the plan. And if you are going to use checklists and something like that all the time, you are sticking too much to the checklist and using too much time for decision-making instead of doing things.
>
> If you only do it in the rational way, you'll never start a search and rescue operation (…). And you have to work intuitively to start up with anything.

Intuitive decisions have the potential to enable decision-makers to act in complex task environments while coping with uncertainty. Intuition allows a decision-maker to preserve the ability to act. Therefore, although many structural conditions push decision-makers toward more deliberation, the intuitive decision mode is accepted in the field if a situation is so complex and demanding that it cannot be properly addressed deliberatively. We continue our discussion of this topic in Chapter 6.

When we asked the experts for their opinion of intuitive decisions in SAR operations, all associated or linked intuition with a decision-maker's experience:

> It takes a lot of time to make a rational decision. Intuitive decisions are very much based on a person's experience.
>
> If you are very well educated and you have a lot of SAR experience, I think you can use the intuitive way. But the rational way is the best way; you have to use these checklists to be sure that nothing will be missing. If you have a very good intuition, you can use it too, but both can be useful.
>
> But you still have to use your intuition to get through it, because you have some experience. I link intuition with experience as well, if you look at it that way. So I would say yes, the conditions, as well. And I think if you put an inexperienced guy used to doing simple tasks out for a big ferry catastrophe off the West Coast of Denmark, for example, they might start to use their intuition, whatever, but if they don't go down and make very fact-oriented rational decisions, they run into garbage and then they will make the wrong decisions.

In our evaluations, most decisions made with a relatively intuitive mode were rated as being of low decision quality. This does not mean that using the intuitive decision mode is substantially inadequate or always leads to poor quality results. Decision-makers switched to intuition only in situations in which complexity was extremely high; therefore, this relationship is not surprising. Still, it underlines the importance of experience for good intuitive decision-making, which might not always be present in highly complex task environments, such as the ones we address in this chapter.

As mentioned in some of the quotations above, decisions are often made with an approach that is somewhere in the middle of the continuum between fully deliberate and fully intuitive decisions. In the interviews as well as when we discussed our research with commanding officers and coxswains from rescue units, we found that an understanding of such a continuum is well-rooted:

> ...rational and intuitive are linked together. It's not one or the other.
> We have all those questions here: We cannot say it is black or white. It is gray, because it depends on the situation, the persons involved, the situation you are put into. So everything is a little gray.
> You have to get the facts right and make rational decisions. But of course there is a lot of intuition in this. So you can't really say it's either this or that. But basically you have to make some really rational decisions.

A tricky thing is that the informational basis on which decisions are made is not always reliable. Not all the necessary information is available right in the beginning of a SAR operation. In addition, not each piece of transmitted information can be verified immediately to be sure that its content is correct. Under these conditions, a decision-maker must gauge the available information that is necessary to properly deliberate as well as the necessity to make the decision within a certain time frame. Considering this problem, we observed quite different approaches in our interviews:

> ... if you have a situation, sit down, have a ciggy, take five minutes, think about the situation instead of just pulling up your gun and shooting two, three times, and do something. Sit back, have a feeling of the situation, use five minutes, 10 minutes, it doesn't matter. But the first 10 minutes are very important just to find a good plan instead of making a plan—a good plan, a bad plan—and then changing it all the time.
> Your experience will give you a plan inside. You will have some picture of the incident. And then you start up. You can't just wait for more information.

Such statements reflect previous insights on the role of decision mode preferences. Although some decision-makers clearly prefer a deliberate mode, others look primarily for intuitive decisions, which guides decision-making.

Decision-making in maritime SAR means continuously reevaluating a situation because each additional fact has the potential to substantially change the general impression. Thus, the planning of measures requires including options for (fundamental) changes.

Checklists and tools may help a decision-maker to decide deliberately. The more a decision-maker is familiar with the handling of procedures and support tools, the more complex situations can be handled in a deliberate manner. If a critical threshold of perceived complexity is exceeded, a switch to more intuitive decision-making ensures the ability to act.

5.1.4.3   About the LICO Concept

Overall, we feel confident to conclude that the concept of LICO is operative and very powerful in coping with major maritime disasters. We made only a very minor critical observation: LICO became the established abbreviation for *local incident coordinator*, used in face-to-face as well as radio communication. Due to the international context of most maritime SAR operations, international parties may be unfamiliar with the abbreviation, as it cannot be found on the official lists of abbreviations and acronyms in the IAMSAR Manual Volume III. In addition, depending on the involved nationalities, their accents, and method of transmission, the abbreviation *LICO* can sound like *leak* or *leakage* if communication quality is poor—a keyword for SAR professionals that can lead them in the wrong direction.

In terms of this and some other aspects, there are still improvements to be made to develop the concept. Yet, as an operational solution to a command-and-control problem, the LICO role is certainly a step forward. From a research perspective, the development of the OSC role provided us with a good opportunity to study how experience shapes approaches to decision-making. This helps us to better understand the broader pattern we found in Chapter 4—that is, that experienced decision-makers adjust their decision modes much less than inexperienced decision-makers—in a practical context.

5.2 DECISION-MAKING IN THE DANGER ZONE: WORKING IN MASS RESCUE OPERATIONS

Decision-making in maritime SAR takes place on many levels: on land, in a rescue coordinator center overseeing several missions at once, and on the operational level in the form of decisions made by the OSC as well as straightforward, hands-on decisions made by a rescue man on scene. We already followed a local incident coordinator's team conducting SAR measures on a large roll-on/roll-off passenger ferry. In this case study, we have a closer look at how decision-making developed aboard a smaller rescue cruiser of 23.3 meters in an international rescue mission in the Aegean Sea. Whereas in the previous parts of this book we focused on the question of which decision mode decision-makers use, the following study complements this perspective by providing insights into how decision routines and expert schemas develop in the first place. To achieve this goal, we follow an international crew situated in a novel and demanding situation of rescuing refugees from overcrowded rubber boats in foreign waters.

5.2.1 INTRODUCTION

The international rescue mission in Greek waters near the Turkish border was a unique research opportunity to obtain a better understanding of organization and team processes aboard a rescue cruiser operating under specific demanding conditions. We observed a ship that was not built for the environmental conditions it had to operate in and was manned by a crew that had almost no experience with the challenging kinds of operations with which they were confronted. The rescue cruiser was constructed to be sailed in the North Sea, with its mudflats and chilly weather conditions. Persons in distress in overcrowded, unseaworthy rubber boats had never been encountered during rescue missions in Northern and Western Europe, where the crew had worked before. Mass rescue operations, such as the ones observed in this case study, not only have a considerable potential to help people in distress. For high reliability organizations such as SAR services, these operations also provide new learning opportunities, for example, handling complex distress situations with multiple persons in the water, which can potentially occur everywhere in the world.

5.2.2 FRAMEWORK

As mentioned in Chapter 3, the mission that guided this research had the focus to build nongovernmental SAR structures in areas faced with a huge number of refugee arrivals from the seaside as well as dramatic fatality rates due to maritime distress. Despite the real deep-set causes of forced migration and refugees, the reasons for maritime distress of refugee boats can be diverse. With a professional view on over-crowded, unseaworthy rubber boats, each SAR worker would judge that situation as one of distress or danger. Nevertheless, many of these boats arrive safely on the shore. However, this is mostly a result of luck.

Crossing the sea in an overcrowded rubber boat is fraught with danger. Some of the boats did not approach the Greek coast due to engine problems; the boats going adrift may capsize or collide with commercial ships or rocks. Some boats begin their journey with a nearly empty gas tank, so they are only able to cross the maritime border between Turkey and Greece but then go adrift and are faced with the same threats described above. The boats that approach the shoreline of one of the Greek islands may come into trouble in the last minutes of the journey. The islands are mostly rockbound; thus, the rubber boats may ultimately crash, leaving their passengers to try to find shelter on the rocks and wait until they are rescued. Many end up with bone fractures, cuts, or injuries from sea urchins.

One of the reasons why boats land on rocks instead of the beach is that the passage usually takes place in darkness around midnight. Smugglers instruct the refugees to head into the lights on the other side of the sea. Most visible lights are sea marks placed in exposed positions—generally on rocks or as markers for areas that pose a danger for shipping traffic. Near the beaches are also dangerous rocks that can lead to accidents.

In an amendment of the SAR convention (International Maritime Organization, 2006) resolution MSC.155(78) adopted on May 20th, 2004, the following sentence was added: "The notion of a person in distress at sea also includes persons in need of assistance who have found refuge on a coast in a remote location within an ocean area inaccessible to any rescue facility other than as provided for in the annex" (ib., Chapter 2.1.1). Therefore, persons stranded on rocks are considered to fall into the field of maritime SAR.

Before the recent influx of refugees began, accidents like this were rare events and concerned one or two unfortunate persons who were saved by a well-trained mountain rescue service. However, with the influx of refugees on the Greek islands, the Hellenic Rescue Team—as the primary mountain rescue service—began to be committed in maritime SAR. As a member of the International Maritime Rescue Federation, the Hellenic Rescue Team was the member to strengthen the joint campaign "Members assisting members." Greek rescue colleagues were always part of the crew on the rescue cruiser that we observed: at least one colleague, but typically up to three. Most were local volunteers who occasionally sailed in operations. That made the crew on the rescue cruiser international and multiprofessional: maritime SAR personnel, lifeguards from whitewater rescue, and experienced volunteers from the mountain rescue.

We report on a mission that began in March 2016 and was centered on the waters around the island of Lesvos. Springtime in the Aegean can be quite cool; and the weather in mid-March was colder than usual that year. Although the air temperatures at the start of operations were between 14°C and 19°C, it decreased to single digits within the following days in the daytime and to just 3°C at night. Locals reported that this was the "coldest winter during the last 20 years." The local infrastructure (buildings) was not made for such weather. The water temperature in the vicinity of Lesvos was around 13°C to 15°C at the time and increased as the year progressed. The weather conditions were variable, depending on the direction of the wind; the topography affects a wind channel between the island and the Turkish mainland. Downslope winds from the east may lead to a calm sea in Turkish waters and rough, choppy waves near the coast of Lesvos. Generally, the weather and resulting sea conditions in this area can change completely within less than one hour.

The mission took place under an uncertain general political situation concerning EU foreign policy as well as the traditionally difficult relationship between the neighboring states of Greece and Turkey. Despite an internationally accepted maritime border between Turkey and Greece, there are different interpretations on both sides concerning the location of the border. As requested from the Greek side, the area of operations was basically limited to the Greek SAR region of responsibility.

Uncertainty was generally a hard player in the mission's framework in various quarters: First and foremost, none of the crew had ever been involved in SAR missions like those that were expected here. Many crew members had worked in international contexts before, for example, in international shipping or trading, as a professional skipper on large leisure crafts, during deployment abroad in the military, or as members of special forces in humanitarian emergencies, such as floods or earthquakes. Notwithstanding the broad range of experiences, the situation that the crew would ultimately face as well as what the mission would demand was uncertain. The home organization strived to provide relevant information regarding the forthcoming mission to prepare the potential crew members as well as possible. Shared experiences and lessons learned from other participating rescue organizations helped to obtain a general idea of the upcoming challenges. Still, a considerable amount of uncertainty remained for each crew member, which increased the perceived complexity of the mission.

Prior to the actual start of the rescue mission, the crew learned a lot from their Swedish colleagues' experiences in the waters of Samos. A main issue concerned critical incident stress management (Mitchell & Everly, 2005) to care for crew members following traumatic events, for example, unsuccessful resuscitations, death of children, or unexpected stress reactions. The purpose of which was to enhance the fact that a rescue operation is, first and foremost, an extreme situation for the persons in distress, who have typically never been in a situation like this and might not even have been at sea before. When all these factors come together, the result may be an indeterminable critical situation during a rescue operation.

5.2.3 OBSERVATIONS

The rescue mission began in the waters off the island of Lesvos and was scheduled for Monday, March 14th, 2016. The rescue cruiser arrived approximately 10 days prior in the Marina of Mytilene, the island capital, and was to be fully equipped and tested during this period. It was planned that some of the small overpass crew as well as a few crew members of the first watch who arrived early would make all the necessary preparations, find local service partners, and make themselves familiar with the area of operations to complete the sea charts. Due to the ongoing tense situation with refugee boats near the coast, the responsible local authorities spontaneously asked whether the rescue cruiser could begin operations one week earlier. The executive director on site decided—based on his own experiences that he made during similar SAR operations near other Greek islands a month earlier—that an early flying start was reasonable with a crew of reduced size and a limited daily timeslot between 600 and 1800 hours.

The author who participated in the mission was scheduled to arrive on March 8th and use the time until the official start of operations for preparations regarding the inquiry. The information about the earlier start was released when the researcher had a stopover at Athens airport. A colleague called from aboard the rescue cruiser: "Are you underway to Greece? Hurry up, your hands are needed! We put out to sea at 5:30 in the morning. See you at the pier and remember the time difference!" Thus, we had to make a flying start with almost no time for preparations aboard. Accordingly, data collection during the first days was limited to a diary, written in the evening, until all crew members had agreed on a procedure for video recordings of the rescue operations.

5.2.3.1 Daily Schedule

A typical day began in the early morning with a crew meeting in the hotel lobby for a short coffee break around five o'clock. If there was something notable to announce or relevant news from the headquarters had arrived late, the coxswain reported that in a short briefing. Afterwards, the entire crew left in a van for the harbor. At the gate to the pier, the Greek colleagues waited to board the rescue cruiser together. All personal belongings were deposited down in the cabins, which were used for multiple purposes, for example, as store and locker rooms. The three cabins were connected by the galley, which contained cooking equipment, a table, and seats and was also used as a mess. Three crew members—those who were designated to work at the interface between the rescue cruiser and the waters as well as the rescue cruiser and the refugee boat—dressed in a survival suit. The other crew members wore the work clothes of their home organization. Everyone carried a walkie-talkie (VHF) that operated on a working channel that could not be interfered with by commercial shipping or leisure crafts.

Before the ship left the harbor, the port authority and the responsible OSC—a Hellenic Coast Guard vessel—was informed that the unit was on call. The OSC assigned a limited area of operations to the unit to search for persons and objects in maritime distress. The first hour was typically dark with poor visibility until the sun rose above the Turkish coastal mountains. Technical equipment supported the search under these poor visibility conditions. The radar, operated by one person and located inside the wheelhouse, was a vital part of the search. The other crewmembers were on the open flybridge and held lookout with binoculars. If the radar operator detected a possible boat in distress, he informed the coxswain on the bridge via radio so that everyone had the information at the same time. Similarly, if someone spotted a potential target, the radar operator was asked to cross-check the suspected position.

When both the visual and the technical search suggested the presence of someone in distress, the coxswain gave the call to prepare for rescue. All crewmembers donned two or three pairs of surgical gloves under their work gloves; the daughter boat was manned with two crewmembers and launched though the aft hatch. The other rescue men and women balanced an inflatable rescue path across the open hatch, placed where the daughter boat was located on board, and inflated it using a CO_2 cylinder. At the same time, the daughter boat—and, if available, a rigid-hulled inflatable boat—manned with two lifeguards approached the refugee boat from two sides and asked if someone on board speaks English. The English-speaking person(s) aboard the refugee boat were requested to translate all given information. Then, the rescue man in the daughter boat explained the next steps of the rescue. The rescue crew threw a rope to the people in the boat in distress, who then fastened it at the outboard engine. Then, the daughter boat towed the rubber boat carefully to stabilize it (see Figure 5.12). Subsequently, the rescue cruiser would head astern to the starboard side of the refugee boat; the rigid-hulled inflatable boat safeguarded the port side in that maneuver. The rescue crew then asserted that everyone was safe and would be rescued if they all followed the instructions. When the daughter boat crew was sure that this was understood, they informed the rescue cruiser via VHF and began the operation as described above.

The rescue cruiser then sailed astern to the starboard side of the refugee boat and fastened docking lines between both. Two lifeguards or rescue men in survival suits safeguarded the end of the rescue path so that no one could jump erratically from the rubber boat toward the maneuvering rescue cruiser. In addition, two persons handled the docking lines. The coxswain or second in command conducted the activities on the afterdeck while the other sailed the cruiser from the flybridge. When the

FIGURE 5.12   Preparation for Evacuation: Daughter Boat Tows Rubber Dinghy.

refugee boat was positioned as safely as possible, the daughter boat came alongside the refugee boat. The persons to be evacuated were instructed by the conductor on the afterdeck to disembark the rubber boat in the following order: children first, then women and parents, and, finally, everyone else. By call of the coxswain, the persons were evacuated in an orderly manner to the rescue cruiser. Luggage was taken to the daughter boat. Children and their mothers were placed in the deckhouse, while the other persons were directed to sit on the foreship.

When all children and their mothers were aboard, a rescue person guided the first adult to the foreship and then instructed the remaining persons to sit there as well. Rescue blankets were handed out; and the people were instructed how to use them. Meanwhile, the persons on board were counted; and the number forwarded to the OSC. The outboard engine of the refugee boat was secured, the empty rubber boat marked "rescued," and then sent adrift. Then, the ship headed back to the port of Mytilene. Underway, aid workers from UNHCR and the International Organization for Migration were informed by phone about the upcoming arrival of migrants in the harbor, the number of persons, and whether special assistance was necessary, for example, when injured or sick persons were on board. The rescue crew always attempted to deliver the luggage to the people on the way back to the harbor or while disembarking. After arriving at the harbor, buses from UNHCR as well as aid workers and physicians from Greek charity organizations welcomed the rescued persons with water, blankets, and dry clothes for the children.

When the first aid ashore was finished, the persons entered the buses and were subsequently transferred to the reception camp in Moria. After disembarking the rescued persons, the entire crew began cleaning up the ship and throwing away any fake life vests and rescue blankets that were left behind. The garbage from the operation was delivered to the pier, where large heaps of life vests already marked the effort of several rescue and border guard units the hours before. After cleanup, the outboard engine from the refugee boat was delivered to the coast guard station on the other pier. One of the crew members then disinfected the inside of the cruiser. When the application time had passed, the coxswain informed the OSC that the cruiser was ready for operation. Usually, the next operation followed closely after. Often, the position of a refugee boat in distress was directly transmitted from the OSC so that the daughter boat crew continued the operation instead of being pulled back aboard the rescue cruiser. Underway to the next case, the crew members ate some chocolate bars, bananas, or other quick snacks. The subsequent rescue operation followed the same scheme as described earlier, so that the rescue unit was eventually back at the pier around 10 o'clock in the morning after finishing two or three operations. Subsequently, the crew began major cleanup of the cruiser—inside and outside—as well as the maintenance of the engines by the engineers. During this time, the coxswain completed his diary with items from the journal.

At noon, one of the crew members brought pizza and other food from a fast food restaurant directly opposite the entrance to the marina. The food was shared in the mess while talking about the impressions of the recent operations as well as ideas about how to improve the employed technologies and procedures. New approaches were immediately tested after lunch—for example, how to handle the rescue path more efficiently or with fewer hands. Others improved the safety equipment, spliced ropes, or explained seamanship to the lifeguards. In the late afternoon, the crew left the cruiser for the hotel, as the debriefing in the hotel began at 1800 hours. After the debriefing, the crew had leisure time, which doubled as standby time, in which the crewmembers were allowed to stay at the accommodation, exercise, or go to a restaurant, but were always to be available on mobile phone if the rescue cruiser was called by the coast guard. The day ended as early as it started with everyone going to bed by 2200 hours.

5.2.3.2   During the Operation

A salient contrast to SAR operations in Middle and Northern Europe concerns the meaning of cues: When we left the harbor the first time, we saw an empty life vest adrift. In Northern Europe, this would be a sufficient trigger to call the maritime rescue coordination center and begin investigations in the vicinity to make sure that there were no persons in distress. There, however, a single life vest in the water meant nothing. Every day we occasionally saw a tube or life vest adrift. Certainly, in the case of a capsized refugee boat, there would be more than one life vest adrift. Thus, normality had completely shifted so that "maritime distress" was always associated with mass rescue operations. No one expected to have a sailing yacht or small fishing boat with one or two persons on board in distress, which is the usual SAR work in the North and Baltic Sea. The huge number of refugee boats in distress also changed the awareness of the rescue workers and gradually made the situation more or less normal. The worst-case scenario, multiple persons in the water, was an everyday occurrence for which the crew was constantly prepared. Procedures were aligned with the medical director beforehand concerning how to practice an on-water triage or the extent to which resuscitation is reasonable in a mass rescue situation. These reflections and the continuous stress situation led to a shift in the definition of normality.

During the first days in operations, we learned a difficult lesson concerning the refugees' luggage (see Figure 5.13). As may be imagined, persons fleeing have all their belongings with them: Their entire life, including things with sentimental value, is limited to one bag. We described earlier how we evacuated the refugee boats, when the persons were evacuated to the rescue cruiser and the luggage was gathered by the daughter boat crew. If we had rescued only one rubber boat in an operation, the daughter boat was pulled back aboard the cruiser, and the bags were delivered to the people while we were heading to harbor. The situation was different when two refugee boats were evacuated in a single operation. In this case, the space on the aft deck—where the daughter boat is located when not in use—was needed for the rescued persons. Due to these circumstances, the luggage was unloaded from the daughter boat to the pier. The crew piled the items of luggage on the pier near the cruiser so that the rescued persons could then go ashore and claim their luggage. Unfortunately, this time the crew members in the daughter boat were so quick that all the luggage was ashore before the persons disembarked the rescue cruiser and thieves—lurking between parked cars and waiting for a lucky chance—stole two items from the pile. The two affected boys who lost their last personal belongings broke down in tears. This scene dismayed all of us, and we agreed that there is urgent need for action to avoid incidents like this in the future.

FIGURE 5.13 Busses Waiting at the Pier; Front: Tubes and (Fake) Life Vests Left behind.

It is no secret that the passage of an overcrowded, unseaworthy rubber boat over sea is not a walk in the park. The choppy Aegean Sea, departure by night from some remote Turkish location, and attacks from multiple sides made the fleeing persons in the rubber boats vulnerable. Many were traumatized by exceptional occurrences, some were injured, and the majority were soaking wet. Especially the children often lost a shoe while sitting with the women down in the middle of the rubber boat. Seasickness and bad air down on the floor made the four- to more than eight-hour passage torturous for the babies, toddlers, women, and the injured. Even if the situation on board a rescue cruiser is not the most comfortable for the rescued, the people felt safe on board; the fear of death slowly decreased and even switched to euphoria for some, such as teenagers throwing their life vests overboard and taking selfies with their smartphones. Some of the rescued persons were dehydrated, hypothermic, and seasick, others were apathetic, and still others talkative or joking. Either way, the time when the refugees were aboard the rescue unit was emotionally charged and had to be acknowledged by the crew regarding safety and predictability.

5.2.3.3 Cooperation with Other Rescue Forces

The sovereign function of maritime SAR in Greece is the Hellenic Coast Guard, which requested assistance and supplied a patrol boat as OSC in the Greek SAR region between Lesvos and the Turkish border. In addition to smaller Hellenic Coast Guard units, other organizations and European authorities were engaged in SAR operations in this area during the first inquiry: A Norwegian rescue cruiser sailing under Frontex (European Border and Coast Guard Agency) command, two smaller but fast Swedish Coast Guard boats, an Italian Coast Guard helicopter, one rigid-hulled inflatable boat from the Hellenic Rescue Team, and a fast Dutch rigid-hulled inflatable boat from a charity organization sailed by volunteers from the Dutch lifeboat organization. At the Northeastern coast of the island, a Spanish charity organization's rigid-hulled inflatable boat was stationed as well as a rigid-hulled inflatable boat from a German charity—both worked exclusively in the far North of Lesvos so that no joint operations were conducted with them.

Due to the unlimited influx of refugee boats, each with 40 to 60 persons on board, mainly the larger rescue units above 20 meters in length conducted the evacuations (see Figure 5.14). The smaller, rigid-hulled inflatable boats safeguarded the detected refugee boats until a larger rescue unit was available for the evacuation. The OSC attempted to involve the rigid-hulled inflatable boats near the coast and the larger units near the border. To make sure that we did not accidentally cross the maritime border, we marked the Turkish interpretation of the boundary line plus a safety margin in our electronic chart display and information system. We discussed this point in advance in our headquarters to come up with an appropriate solution as well as what is to be done if a boat is in distress in Turkish waters. We agreed that situations like this would be conducted according to international standard procedures and, apart from that, to follow the instructions from the OSC.

Although our unit began daily operations before sunrise and worked regularly, the Frontex units (European Border Defense Mission) pounded the beat unpredictably and often during the night, searching also for smugglers. At this point, it is necessary to differentiate between the different tasks that a coast guard has besides SAR: There are many more official duties concerning law enforcement for them to fulfill that cannot be done by a SAR unit or which may end in a tragedy if a rescue unit attempts to rescue armed criminals by accident. A few weeks before our mission began, another rescue unit was attacked with guns by smugglers who, contrary to the rescuers expectations, were not in distress. To avoid a repeat of this incident, the

FIGURE 5.14   Head Count during the Transfer to the Harbor.

OSC only delegated tasks when they were sure that the object in distress was really a refugee boat.

We conducted all operations appropriate to our position in the command system. Collaboration with the other charity organizations working in the maritime vicinity of Mytilene was also professional and amicable. We conducted exercises together, shared procedures and experiences, and exchanged phone numbers. Near the end of our mission—during the second inquiry in May 2016—joint search and rescue exercises with other charities that were less experienced in the maritime domain became more important. This is because some of the experienced organizations downscaled their volunteer engagement; the resulting gap could only be offset by newcomers. In addition to the final trainings with our Greek partner organization, the Hellenic Rescue Team, came other national charities that needed basic training from us as well as from a Canadian rescue organization that delegates instructors to train these charities' volunteers. Our mission ended after three months in June 2016.

5.2.3.4 Team Processes

In the next step, we will examine how team processes shaped the development of action routines. The description of the general evacuation procedure conducted in this unique kind of mass rescue operation has already shown that the method in each operation was of a similar type. It was obvious that, due to the structural capabilities of the rescue cruiser, an evacuation to the rescue cruiser by the afterdeck via the rescue path is the safest practice. Nevertheless, after each operation, we identified weaknesses in the procedures and/or gear that had to be improved. The crew spent hours every afternoon improving the methods, making them more reliable, safer, and easier to handle. Striving for continuous improvements meant observing an iterative process. This can be exemplified by the method of throwing a line from the daughter boat to the refugee boat. Typically, when other boats and ships are adrift and require towing assistance, a simple rope gets thrown to the adrift boat and is knotted somewhere to be towed by the cruiser or a daughter boat (see Figure 5.15). To conduct this basic procedure, it is necessary that someone is aboard the adrift boat who possesses basic skills in seamanship, for example, the ability to tie a proper knot. The overcrowded rubber boats have just one part of the hull that is steadier than the rest: the transom where the outboard engine is fixed. Because it is very unlikely that a person with basic seamanship skills is aboard the refugee boat, we decided to place a carabiner hook at the end of the line. This would enable a layman to make a proper line connection to the daughter boat.

However, we soon had serious concerns regarding the safety of throwing a line with a heavy steel carabiner hook toward a boat with babies and toddlers sitting down on the floor—it would be irremediable if one of them got hit by the heavy hook. Thus, we decided to instruct all newly arriving crewmembers to throw the hook as far back in the rubber boat as possible for safety.

With the next crew, the first lifeguards from the whitewater rescue arrived. When we introduced the procedure that we used to prepare the evacuation, they asked whether we had throw bags. No one was aware of this solution before, and thus we added a throw bag to the towing lines in the daughter boat as well as to the lines aboard the cruiser that had to be thrown to the refugee boat. All the improvement processes were done step by step; ideas were negotiated and discussed among crew members and, if there were two

FIGURE 5.15 Before the Evacuation from Refugees' Perspective.

competing ideas, eventually decided by the coxswain. He pushed the improvement forward often saying, "Let's have a look at... I'm not happy with the way we do this..." The similarity of the operations provided the unique opportunity to test new methods and gear in practice the following day to decide whether it was practical (see Figure 5.16). If necessary, there was an entire afternoon in which to improve the approach as well as the next day to test the latest improvement. This process of incremental, feedback-driven learning strongly drove the development of practicable solutions.

The roles and hierarchy on board were somewhat similar to the structures at home. The overall responsibility was with the coxswain, an experienced SAR professional. Second in command was also a professional, working as a coxswain at home, who was familiar with the cruiser class and equally qualified as the responsible coxswain. The engine and all related equipment operated under the responsibility of the chief engineer, who had the same professional position at home. The other rescue workers on board were mainly volunteers from the maritime SAR service as well as volunteers from the whitewater rescue. Between the rescue workers, there was no hierarchical difference. Depending on the experiences and skills, everyone campaigned their own strong points and were available as needed. During the operation, each person knew what to do for the position at which he or she was working. All this took place without much talking with each other. Crew members thus already possessed or developed expert schemas and action routines that could guide intuitive decision-making in the regular ship handling and SAR tasks. Only the conductor on the afterdeck—the second in command or the coxswain—intervened if they perceived an inconsistency and, therefore, acted as a safeguard for false decision-making.

The working focus of all rescue workers was on the refugee boat. In subsequent interviews, we asked the crewmembers to remember the rescue operation: What did you focus on? The answers were nearly all the same. All mentioned the danger that the fragile, poor quality rubber boats could be damaged by the rescue operation,

FIGURE 5.16 Test of Improvements on the Rescue Equipment (Lowered Rescue Path).

resulting in multiple persons in the water from one moment to the next. This worst-case scenario induced by the rescue activities was the shared nightmare of the crew members, and each stove to avoid circumstances that could end up that way. Depending on the individual point of view during the operation, the colleagues had understandable beliefs about what the worst case would look like. For example, "If we don't tow carefully enough, the transom of the boat tears off, both tubes slap together, the floor folds away and the children, mothers, and babies are goners."

5.2.3.5 Second Inquiry at the End of the Mission

On April 3rd, after more than three weeks of service, the participating author left the crew to return two and a half months later for a second evaluation with the goal of ascertaining any changes. During that time, fundamental changes had occurred: first and foremost on the political level. The EU-Turkey Refugee Deal entered into effect one day after we finished the first inquiry. As a result, the continued combination of coast guard forces in Turkish waters that we observed during the last days of the first inquiry as well as the presence of NATO warships patrolling mainly in Turkish waters meant that the number of arriving refugee boats severely decreased. At the end of May, one or two boats infrequently arrived in Greek waters at night—sometimes even none. The area was tightly controlled and monitored by NATO ships from the Turkish side as well as an offshore supply vessel under Frontex command from the Greek side.

Meanwhile, our rescue cruiser continued the daily patrolling. Yet, it now started its day much earlier in the morning. The briefing took place at 2:00 in the morning so that the cruiser was manned by 2:30. By that time, the cruiser had forward-looking infra-red devices to support searches. The division of labor between the colleagues search-ing with binoculars and infrared from the flybridge and another colleague operating the radar in the wheelhouse was the same as previously. When we located a potential target, the OSC was informed by VHF radio, and the crew then waited for further instructions. Other units almost always had monitored and clarified the target already.

Also, the strategy of the OSC had changed between the two inquiries. If a refugee boat was detected, one or two very fast units were sent to the boat. Then, the boat was guided by the authority units to the port of Mytilene; no one touched the rubber boat as long as it was able to sail. With this strategy, the smaller but much faster Greek and Swedish coast guard units, which did not have the capabilities to evacuate a refugee boat, were optimally deployed. This meant that we could focus on patrolling, helping to verify sightings, standing by for safeguard if an evacuation was necessary, and finally, conducting trainings and exercises with other charity organizations. Thus, we attempted to practice the evacuation procedure every day, which had been fine-tuned since the first inquiry (and was called the "Mytilene maneuver" by some).

We again interviewed crew members and asked them what they focused their thoughts on. Some reported about their motivation to be a volunteer in this mission, others of very hands-on matters. Some reflected on the situation in general—how awful the situation in Syria, Iraq, and other regions must be to risk such a dangerous passage with their children—and thought about the circumstances under which they would take this step with their own children. In contrast to the first inquiry, we did not observe many similar ideas and perceptions in the crews operating at the end of the mission, although the procedures were still practiced.

Furthermore, an additional procedure was practiced to evacuate persons from rocks. The daily trainings were also used to perfect the collaboration between the lifeguards, their inflatable rescue boats, and the operations of the rescue cruiser. Once a week, an interorganizational search and rescue exercise was conducted with the rigid-hulled inflatable boat crews from the other charity organizations. Our instructors developed and conducted this exercise in cooperation with Canadian SAR instructors. The focus of these exercises was, for example, standard procedures in a search with multiple units.

5.2.4 Resume

5.2.4.1 Decision-Making as Anticipation

This case study provides some quite fascinating aspects. The mission contained sub-stantial uncertainty for all participants, placing them in decision situations for which no action scripts were available. The mission also had a variety of other stressors: Will we have multiple persons in the water who risk drowning? Is there a risk for the rescue workers to contract a serious disease? Thus, this mission was a unique oppor-tunity to see action scripts and shared mental models develop.

The overall uncertainty was countered by decision-making in advance—both before the start of the mission and between the operations—very deliberate and analytical. Possible problems and variants were anticipated and modeled as well as solutions or procedures arranged. Influencing variables were collected and discussed by the team to come up with plausible and practicable solutions. Deliberate decision-making in advance helped the crew in a rescue situation to have deliberately created solutions at hand, for situations for which no action scripts had existed before. This helped the crew members to not become overwhelmed by the complexity of the situation. Mental modeling and deliberating in anticipation of future decisions is an interesting complement to the naturalistic decision-making perspectives on intuition in action (see Chapters 2, 4, and 6): In situations in which expert decision-makers had no expert schemas in place, they attempted to consciously model likely scenarios and potential solutions to improve actionability in time-critical situations. This decoupling of decision and action is an interesting method to avoid the bottleneck problem that is characteristic of deliberate decision-making.

5.2.4.2 Standardization of Procedures through Incremental Learning

The rescue cruiser crew used all opportunities to create procedures for the new kind of operation—the evacuation of overcrowded, unseaworthy rubber boats as well as the case of evacuating persons stranded on coastal rocks—which they did not have before this mission began. These procedures were continuously refined daily and actively introduced and forwarded to new arriving crew members. The process of standardization itself was based on recent experiences, various attempts, deliberative decisions, and testing of practicability. From the evacuations in the beginning of the mission to the last operations, it was possible to continuously decrease the time necessary for the evacuation—up to half of the initial time. This process of incremental learning allowed a shared development of expert schemas and optimized action scripts that could be used to fruitfully employ relatively fast, intuitive decision modes during the actual assignment.

Emotion management also played an important role in the development of the crew's capabilities. The crew was completely aware that emotions can affect the (ad hoc) solutions of intuitive decision-making. That is why different scenarios that could potentially occur were discussed in advance to provide rescue workers with a set of possible solutions and actions if one of these highly complex situations were to occur. The worst-case scenario was present in all minds and was as such a type of common picture. The common picture could have been created via the crew's discussions about improvements or extrapolated experiences coupled with a domain-typical disposition to prepare for the worst possible situation. During the first inquiry period, we found the shared mental model under conditions of several rescue operations per day and less developed standard operating procedures. In the second inquiry, when the procedures had been perfected but the frequency of rescue operations had decreased to zero, there was no longer such a clear mental focus. Learning was no longer that important, and the exposure to high-stress situations was less of an issue. Accordingly, the crew no longer had such a strong unifying focus.

5.3 METHODOLOGICAL REMARKS

In the conclusion of this chapter, it is helpful to offer some methodological comments, starting with Study 1: First, when comparing the decision-making of Team 1 and Team 2, we saw that Team 2 ultimately performed better. However, it is important to note important background conditions. In 2014, the concept of the LICO was new. Thus, the associated roles and responsibilities were still in development. Between 2014 and 2015, partially based on the experience made during the 2014 exercise, the concept was refined, and some of these lessons were taught in a workshop before the second inquiry. Team 2, therefore, had a substantially better starting position in terms of cognitive preparedness. In addition, there is also a certain danger of falling prey to outcome and hindsight bias when judging decisions. We and the experts involved in the interviews, strove to interpret the decision situations "...given the information available at the time it was made..." However, it is of course not possible for any observer to exactly replicate the knowledge that a decision-maker had at a particular point in time. Thus, the effect of such bias cannot be ruled out completely. In consequence, we urge caution when judging the performance of decision-makers in this case study.

In Study 2, we limited video capture and provided veto rights to the involved rescue workers due to ethical concerns regarding filming refugees in emergency situations and rescue workers in potentially traumatic and stressful situations. We used action research as the method of choice for Study 2, which meant that missing data in the research diary as a result of fatigue or "going native" bias might be an issue.

Still, we are sure that the video-based approach we used provided us with excellent, in-depth insights into decision-making in complex task environments.

REFERENCES

International Maritime Organization. 2006. *SAR convention. International convention on maritime search and rescue*, 1979; as amended by resolutions MSC.70(69) and MSC.155(78); 2006 edition. 3. ed. London: IMO.

International Maritime Organization. 2014. *SOLAS. Consolidated text of the International Convention for the Safety of Life at Sea*, 1974, and its protocol of 1988: Articles, annexes and certificates; incorporating all amendments in effect from July 1, 2014. 6. ed., consolidated ed. London: IMO.

International Maritime Organization; International Civil Aviation Organization. 2010a. *Mobile facilities*. 2010 edition. 8th ed., incorporating amendments through 2009. London: IMO (IMO publication, IAMSAR manual/IMO; ICAO; Vol. 3).

International Maritime Organization; International Civil Aviation Organization. 2010b. *Organization and management*. 8th ed., incorporating amendments through 2009. London: IMO (IMO publication, IAMSAR manual/IMO; ICAO; Vol. 1).

International Maritime Organization; International Civil Aviation Organization. 2013. *Mission co-ordination*. 2013 ed. 6. ed. London: IMO (IMO publication, IAMSAR manual/IMO; ICAO; Vol. 2).

Lübcke, T. 2016. Videography in dynamic environments. Restriction on and opportunities for multiplicative GoPro camera usage. Videography and interpretative videoanalysis. *9th International Conference on Social Science Methodology*. Leicester, 9/14/2016.

McDonald, S. 2005. Studying actions in context: A qualitative shadowing method for organizational research. *Qualitative Research*, 5(4): 455–473.

Mitchell, J. T., & Everly, G. S. 2005. Critical incident stress management. *Handbuch Einsatznachsorge*. German edition. 2., Edewecht: Stumpf & Kossendey.

Møller, A. 2014. *Guide to coordination of major SAR incidents at sea. Mass rescue operations*. Edited by Naval Warfare Centre. Frederikshavn. Available online at http://pre2016 .balticsarex.org/images/documents/sarex2015/Guide_to_Coordination_of_major _SAR_incidents.pdf, checked on 9/19/2016.

6 Insights, Learnings, Recommendations, and Paths Forward

6.1 SUMMARIZING THE KEY INSIGHTS

6.1.1 OUR UNDERSTANDING OF DECISION-MAKING IN COMPLEX TASK ENVIRONMENTS

At the outset of this book, we introduced a variety of questions: How do decision-makers in complex task environments approach difficult and not-so-difficult judgmental decisions? What is the role of experience as well as differing degrees of situational complexity? What has all this to do with decision quality and the ability of employees in complex environments to successfully handle their oftentimes dangerous and almost always important tasks? How are decisions shaped by organizational structures? We took an in-depth look at the current state of knowledge regarding these questions in Chapter 2, which helped us understand the advantages and disadvantages of the different decision modes. However, we saw that previous research left a variety of gaps in our understanding of decision modes in complex task environments unaddressed. We then provided empirical evidence to answer those questions in Chapters 3 through 5. Now, it is time to summarize what we have learned and provide answers to these central questions that inspired us to conduct our research and, in the end, write this book.

Before we do so, it is important to once again mention the conceptual foundation of the book because, as a basis for our understanding of decision-making and decision modes, we must understand the processes that govern human cognition and, ultimately, human decision-making. In particular, it is important to understand that human cognition consists of two different systems: Type 1, which is an automated, subconscious, and holistic response to decision situations, and type 2, which governs conscious, deliberate reasoning and analysis. Type 1 is effortless; when decision-makers apply type 1 processing, they are often not even aware that they made a decision at all. Instead, they act on what they perceive to be "natural" or "the only reasonable thing to do." The feeling of certainty that sparks such evaluations is an important part of type 1 decision-making. Yet, type 1 solutions do not always come easily. Thus, for this and other reasons, decision-makers might be inclined to consciously think about a decision—that is, apply type 2 processing, which is effortful and much slower than type 1. Decision-makers cannot do much else while they deliberate on a decision problem. Still, type 2 reasoning also has powerful advantages.

In particular, it allows for abstraction, the application of consciously developed or selected rules, and the selective inclusion or exclusion of information in a decision process (Dane & Pratt, 2007).

These two systems are cognitively distinct (Evans, 2006). Type 1 reasoning is always activated when a decision must be made even if the decision-maker is not consciously aware of the decision situation. This is not true for type 2 reasoning, however. Instead, it is in the discretion of the decision-maker to use more or less deliberation on a specific decision, which then replaces subconscious decision-making to a certain degree. This mechanism of gradually selecting the degree of deliberation put into a decision (or the gradual reliance on intuition to frame it the other way around) leads to a continuum of what we called decision modes in this book. One extreme of the continuum is purely intuitive decision-making, a situation in which decision-makers apply no deliberation at all but, instead, rely entirely on their type 1 processing. Examples of this type of decision are highly routinized, everyday decision situations, such as driving a car, but also decisions based on pure and instantaneous guesswork. The first example is mature intuition—that is, intuition based on sound experience and well-established expert schemas. The second example is immature intuition, in which a decision-maker has no real understanding of a decision situation but simply chooses something without a basis to evaluate the relevance or appropriateness of the choice (Baylor, 2001). It is not surprising that mature intuition tends to be a powerful decision tool, whereas immature intuition tends to be an often less effective mental shortcut (Kahneman & Klein, 2009).

The other extreme on the continuum of decision modes are decisions in which the initial intuitive choice is completely overruled by deliberate reasoning. That is, decisions for which a decision-maker builds his or her decision solely on the results of analytical, conscious processes. This is the case when, for example, a decision-maker consciously evaluates and weighs all possible options, carefully selects a decision rule to choose between these options, consciously selects which information to use in the decision, and then applies this algorithm to process the selected information to arrive at a decision.

Hence, the dual-processing perspective on human cognition and the freedom decision-makers have over the degree of deliberation they put into a decision results in a continuum of decision modes, ranging from fully intuitive to fully deliberate. We argued at the outset of the book that we expect that decisions are very rarely purely intuitive or purely deliberate. Instead, we suggested that decision-makers likely apply some degree of deliberation to most decisions but that this degree varies. We set out to test this assumption and establish, in particular, how experience and varying degrees of situational complexity affect the choice of decision mode. These are the guiding questions of this book; and now it is the time to answer them.

6.1.2 Answering the Central Questions of the Book

Let us begin with the initial guiding question: How do decision-makers actually make decisions in action?

Previous knowledge regarding the choice of decision mode in complex task environments is relatively scarce and also one-sided. From laboratory experiments, we can learn a lot about the cognitive mechanisms involved in making judgmental decisions as well as about isolated mechanisms affecting decision-making and decision quality. Unfortunately, however, we can learn little about decision-making in real life, in which stakes are high—that is, there is danger for life, health, the environment, or the property of the decision-maker or others who are affected by a decision—and in which conditions that influence a decision situation are overlapping and oftentimes causally ambiguous (Lipshitz, Klein, Orasanu, & Salas, 2001a). The empirical evidence from previous research on real-life decision-making, namely the "naturalistic decision-making" stream, tends to suggest that decision-makers approach judgmental decisions in complex, real-life environments with a predominantly intuitive decision mode—an insight developed in interview-based studies on decisions made under extreme time pressure (Klein, Calderwood, & Clinton-Cirocco, 2010). We, in turn, set out with the assumption that this perspective is a simplification and that proficient decision-makers will, instead, vary in their approach to complex decisions.

Confirming our expectation and contradicting the predictions of the naturalistic decision-making stream, we see that decision-makers do indeed vary quite substantially in their choice of decision mode. In particular, we see that a majority of decisions include moderate amounts of deliberation, whereas full intuition and full deliberation are rather rare. These results substantially extend previous knowledge. Contrasting the "naturalistic decision-making" approach, we showed that decision-makers generally use quite a lot of deliberation in complex task environments. Still, taking a closer look, this apparent contradiction can be resolved relatively easily if we consider an important contingency condition: time pressure. The focus of the classic studies of the "naturalistic decision-making" stream were decisions under extreme time pressure. These decisions will, quite naturally, be made with an intuitive decision mode, as extremely high time pressure makes it virtually impossible to use deliberation, which is comparitively slow. If you are in a burning building and the roof is collapsing, you cannot deliberate on your options but instead have to rely on a fast mental simulation when deciding if you have to get out or can continue with attempts to extinguish the fire. However, we argue and show for our field that such time pressure is, for many decision-makers, quite rare. We examined decision-making in maritime SAR, where decision-makers typically have at least some time to make their decisions—that is, there is some freedom to deliberate. Even with multiple persons in the water, ships burning, or many actors requiring coordination, decision-makers typically have several seconds or even minutes to think through different options in a particular situation. Still, taking time for deliberation might have negative consequences, as the decision-maker cannot do anything else during this time. Still, there is a real choice regarding decision mode, which is not the case in situations in which time pressure is very high. We thus conclude that "naturalistic decision-making" studies a very specific decision situation that does not occur very often in many real-life decision environments and, thus, cannot really inform us on the question of which decision mode decision-makers actually rely on in complex task environments. Instead, we learn that a broad variety of decision modes are used.

After establishing this initial answer, the logical follow-up question is that if decision modes vary substantially, what affects which decision mode a decision-maker actually uses? We broke this question down into three smaller ones and will begin with the following: Is the decision mode determined by characteristics of the decision-maker?

We know from laboratory research (e.g., Betsch, 2008) that decision-makers have a stable preference for a more deliberate or intuitive decision-making mode. This raises the suspicion that this preference could determine how decision-makers approach decisions in real-life, complex task environments, which could explain the variance in decision modes we observed.

Across our empirical studies, we found that the effect of decision mode preference was relatively limited. An effect is certainly there, as some studies showed, but in no way dominates the choice of decision mode. Instead, we found a strong effect of characteristics of the situation, measured as subjective situational complexity. We will discuss this in the following. We also did not find that other decision-maker characteristics, such as stress resistance, mindfulness, age, or experience, dominated the choice of decision mode, although experience, in particular, has important effects, which we will discuss later.

The answer is therefore that both matter: characteristics of the situation as well as characteristics of the decision-maker. In addition, the characteristics of the decision-maker influence but do not determine the choice of decision mode.

Having established that situational characteristics matter, the next question in need of an answer is then how this happens: How does the perception of complexity in a situation affect the degree to which decision-makers rely on deliberation or, instead, trust their intuition?

Situations can vary substantially in terms of complexity. Uncertainty might be high or low; there might be enough, too few, or too many cues to consider; resources might be scarce or generously available; and the situational conditions might or might not affect the abilities of a decision-maker to effectively make decisions. As we saw in the introductory story presented in Chapter 1, a decision that would not be very challenging during daytime and calm sea conditions can be dauntingly complex during a stormy night with a large amount of potentially confusing or conflicting information and a substantial lack of resources. At the outset, however, it was not clear how these variations in complexity would affect the choice of decision mode. Would an increase in complexity indicate a need for more deliberation, as research on metacognition suggests (Alter, Oppenheimer, Epley, & Eyre, 2007; Thompson et al., 2013)? Or would high complexity lead decision-makers to trust their gut feelings more, since the benefits of the intuitive mode are particularly relevant in highly complex situations, as naturalistic decision-making research implies (Klein, 1995; Lipshitz, Klein, Orasanu, & Salas, 2001b; Nemeth & Klein, 2011)?

The answer we found is that the relationship between situational complexity and the use of a deliberate decision mode is inverse U-shaped. In the low-complexity range, decisions tend to be made with an intuitive approach—a finding that fully aligns with our understanding of routine-based intuitions as quick solutions to low-complexity problems (Baylor, 2001). As complexity increases, decision-makers use more and more deliberation. It seems that subjectively perceived increases in complexity do indeed act as a trigger for more conscious effort, as the concept of metacognition suggests (Alter et al., 2007). Yet this result might also relate to organizational pressure to use deliberation, which we will discuss later. Eventually, the use of deliberation is at a maximum; and as decision complexity increases even more, decision-makers begin to reduce their use of deliberation once more. At the upper end of the complexity spectrum, we then see a situation that resembles the predictions of naturalistic decision-making: Decision-makers predominantly use intuition with only a fraction of deliberate decision-making remaining. This finding does not contradict previous research. Instead, it resolves previous contradictions. We now understand more clearly how the tendency of complexity to trigger deliberation as well as the benefits of intuition for highly complex decisions come together to influence decision mode selection.

We also know that experience plays a key role in determining the effectiveness of decision-making in both an intuitive and a deliberate mode. Yet we still did not know *how experience relates to the choice of decision mode.* We will answer this question in the following.

Again, previous conceptual and empirical insights are rather inconsistent. On the one hand, researchers suggest that experience benefits intuitive decision-making to such a strong degree that experienced decision-makers should be reluctant to use a different decision mode, considering the many advantages of the intuitive mode, such as holistic processing, high speed, and low cognitive workload (Klein, Orasanu, Calderwood, & Zsambok, 1995). This perspective would imply that experienced decision-makers consistently use more intuition than less experienced decision-makers. On the other hand, others voiced concerns that experience might lead decision-makers to lose cognitive flexibility—that is, to lock in to established patterns and tried-and-true approaches (Dane, 2010). This would suggest that decision-makers stick to one particular decision mode, irrespective of which mode that is, and do not adjust their decision mode to situational demands—a situation that has been called "cognitive entrenchment."

Regarding the role of experience in the choice of decision modes, we found rather strong support for the cognitive entrenchment hypothesis. We saw that experienced decision-makers did indeed adjust their decision mode less than less experienced decision-makers based on changes in situational complexity. In addition, in one study, we found that there was no change at all. One of the in-depth case studies, in which we followed an experienced and a less experienced decision-maker through a highly demanding training scenario, also showed that the experienced decision-maker took care to set the stage such that it enabled him to apply a particular decision mode consistently throughout the exercise, whereas the less experienced decision-maker did no such thing but, instead, adjusted to the perceived situational demands. These findings strongly support the notion of Dane (2010) that cognitive entrenchment plays an important role. Still, we do not know if this cognitive lock-in, which could also be understood as a form of cognitive stability or resiliency, is necessarily negative, as previous research on cognitive entrenchment implies, although there are some indications in that direction. Specifically, the data showed that highly experienced decision-makers are not significantly more satisfied with their decisions than less experienced decision-makers are with theirs. This is surprising, as we know that experience strongly benefits intuitive as well as deliberate decision-making (see Sections 2.3.3 and 2.4.3). One explanation (see Section 4.3.5 for others) for this finding is that, due to their lack of adjustment of decision modes, experienced decision-makers do not reap the full benefit that a high level of experience should imply in terms of decision-making performance. Still, it is beyond the focus of this book to fully address this question.

Finally, we looked at the organizational side: Do the organizations we study prefer or support a particular decision mode, and does this matter for the actual choice of decision modes?

Decision-makers do not act in isolation but instead are embedded in a certain organizational environment, which provides rules on how decisions should be approached. These rules can be explicit, such as written-down guidelines, or implicit, such as unarticulated expectations that a particular behavior is required or encouraged. Both can strongly affect the performance of employees in organizations, in particular in safety-oriented environments (Boin & Schulman, 2008; Hollnagel, 2006).

In the case of maritime SAR, we found that the organizations encourage and partially expect deliberate approaches to decision-making. This expectation is partially explicit, as communicated in trainings, and partially implicit—enforced, for example, through a heavy focus on debriefings in which decision-makers must justify their decisions in front of others. As intuitive decisions are subconscious and the applied decision rules typically unknown to the decision-maker, intuitive decision-makers find it much more difficult to defend their decisions in debriefing situations. This is consistent with our observation that a preference for intuition relates to strong stress resistance. Overall, there is a substantial pressure toward deliberation, which might, to some degree, explain the pattern we observed regarding decision-makers' responses to increases in subjective situational complexity. In this field, decision-makers are "entitled" to use intuition only in low-complexity decisions and in decisions when there is not enough information on which to base a deliberate decision. Highly experienced decision-makers are licensed to use more intuition due to their superior experience. A comparable observation that the "right" to use intuition must be earned has been reported by other researchers (Hensman & Sadler-Smith, 2011). We will discuss the implications of these observations later. This organizational pressure might also be an explanation for the observation that experienced decision-makers do not adjust their decision modes much in accordance with situational complexity. Due to their experience, they are more free than less experienced individuals to choose and therefore need less justification from the specific environment to use a particular mode. Still, we see that it is not the case that experienced decision-makers use more intuition than less experienced decision-makers. Thus, this explanation cannot fully explain the empirical pattern we observed.

In sum, we believe that we are now in a strong position to conclude that three conditions, in particular, drive the choice of decision mode:

- Individual preferences for a particular mode
- Perceived complexity of a situation
- Organizational environment

These conditions interact to some degree. Specifically, we see that the organizational environment affects how decision-makers experience the need implied by changes in perceived complexity. However, they are also independent effects—that is, none of the effects dominates the others. Experience is an important moderating condition. These effects produce an inverse U-shaped relationship regarding the relationship between perceived complexity and the applied decision mode for less experienced decision-makers, whereas highly experienced decision-makers do not adjust their decision mode significantly.

These results provide new insights into how decision-makers in complex task environments approach decisions contingent on the key characteristics of decision mode preference, subjective situational complexity, and experience, advancing our knowledge regarding the "use factors" (Dane & Pratt, 2007) of intuition and, in consequence, also deliberation in complex task environments.

6.2   WHEN SHOULD DECISION-MAKERS APPLY INTUITION
AND DELIBERATION IN COMPLEX TASK ENVIRONMENTS?

Now that we have established when decision-makers use intuition and deliberation, there is a complementary question: When *should* decision-makers actually rely on intuition or deliberation? Or, put differently, is the inverse U-shaped relationship we observed an effective approach to cope with differing degrees of situational complexity, or is it a problem that decision-makers and trainers should try to overcome?

In Chapter 2, we discussed the advantages and disadvantages of intuitive and deliberate decision-making. This discussion helps us to now develop an informed answer to this question. Let us first recollect a crucial distinction at the outset of the discussion: the separation between two different types or concepts of intuition— mature versus immature intuition (Baylor, 2001). Immature intuition is a situation in which decision-makers apply intuitive judgments as mental shortcuts in complex tasks without the necessary level of expertise to fully understand the intricacies of the related decisions. Immature intuition resembles guesswork and typically falls prey to a variety of judgmental biases, especially attribution substitution, with which parts of an originally complex decision are replaced by simplifications that allow easy solutions that tend to be highly inaccurate (Kahneman & Frederick, 2002; Tversky & Kahneman, 2000). Mature intuition is not free of biases either, but has a critical advantage in that it makes full use of the benefits of type 1 processing—that is, subconscious holistic information processing and decision-making. In the context we examined, decision-makers are well educated and well trained professionals. In most situations, they are therefore able to apply mature intuition, if they choose to do so. The advantages of mature intuition can be roughly grouped into two categories: cognitive efficiency and holism. Cognitive efficiency stems from the high speed and low cognitive workload of intuitive decision-making, whereas the holism advantage rests on intuition's ability to draw from contextualized (experience-based) knowledge that is only fully available on a subconscious level. Deliberation, on the other hand, has its large advantage in the ability to choose decision rules and the information that should go into a decision as well as the disadvantages of suffering from slow speed and high cognitive load.

For low-complexity decisions, the efficiency benefits of intuition are clearly visible, while, at the same time, the need for selective cue processing and purposeful rule selection should be quite low. Low-complexity decisions, in our understanding, are decisions for which cues are easily perceivable (e.g., due to favorable weather conditions) and for which uncertainty is limited and sufficient resources are available. Intuition provides a very effective solution to this type of situation. Thus, the heavy use of intuition for low-complexity decisions is to be applauded from a research point of view. Accordingly, the policy of the SAR organizations to encourage intuition in low-complexity decisions is well placed.

These relatively clear optimality criteria vanish, however, as complexity increases. As cues become more difficult to perceive and uncertainty increases, the benefits of both the holistic processing of an intuitive mode and the analytical power of a deliberate mode become more important, whereas efficiency is, from a relative perspective, no longer centrally important, as managing the greater complexity then becomes the focus of decision-making. Based on the suggestions of naturalistic decision-making, recognition-primed approaches, which are more intuitive than deliberate, should be a powerful tool for moderately complex decisions. On the other hand, it is exactly in this range of complexity where deliberation shines most. Complexity, here, is such that the cognitive load is not overwhelming, whereas uncertainty and an increased difficulty obtaining and processing relevant information make much use of the advantages of deliberation, such as including outside and unrelated information or purposefully selecting appropriate decision rules. Thus, we would conclude that, for the midrange of complexity, deliberation, which is being used most heavily, is a good choice, whereas intuition would be a solid option, as well. Accordingly, it makes sense from an organizational perspective to encourage deliberate decision-making in this type of decision situation, although this would not be mandatory.

However, judging the highest levels of complexity is rather difficult. We observed a shift toward more intuition in that respect, which perhaps does not reflect the ideal approach. An indication is that satisfaction with a decision heavily declines with an increase in complexity. Although this is to be expected, it also indicates that the return to intuition at the outer end of the complexity spectrum does not remedy the problems associated with high complexity. Deliberation is certainly difficult under these circumstances when cues are difficult to perceive and process, uncertainty is high, and resources are stretched to the limit. Thus, there is good reason to return to intuition. Still, there is one important consideration to be kept in mind: High-complexity decisions occur very seldom in real life—a fact we learned when attempting to collect data from real-life assignments (see Chapter 3) and which was confirmed in numerous formal and informal interviews. Real-life experience tends, to an overwhelming degree, to relate to low- and middle-complexity decisions. As a result, even proficient decision-makers will often not develop the appropriate expert schemas to deal with high-complexity decisions efficiently using a mature intuitive mode. The introductory example from Chapter 1 provides an illustration for this argument. The captain of the vessel in distress had to make a quite difficult decision: Should he lead his crew to the dangerous forecastle for possible evacuation or stay in the safe deckhouse and forego the chance for quick evacuation? In heavy weather, with a variety of partially inconsistent cues and little indication of the probability of another heavy wave hitting the ship at any time, this is clearly a highly complex decision, and it is a decision embedded in a situation the captain very likely never encountered previously. Having to make a decision with a lack of experience in comparable situations is a bad starting point for intuition, as intuition is, in this case, likely to degrade into immature intuition—that is, it will be largely guesswork. For these situations, deliberate analysis is the better course of action. Deliberation reduces the probability that simple mental shortcuts govern the outcome of a decision process and allows the decision-maker to consciously select rules and acquire additional inputs if the basis for a good decision is not there. The high complexity will probably imply that a decision-maker's ability to fully use complex analytical algorithms is substantially limited. Yet, in any case, he or she would be able to consciously select a decision rule instead of relying on immature intuition.

Therefore, seeing how rare very complex decisions are for decision-makers, it seems advisable to use deliberation as much as possible for this type of decision. Of course, deliberation also does not provide a tool to overcome chance in decisions that involve a strong element of uncertainty. However, it might help decision-makers to think about probabilities clearly—an element in which intuition is notoriously ineffective (Tversky & Kahneman, 1979).

In sum, we see that the choice of decision modes, as we observed for less experienced decision-makers, is oftentimes quite close to what we would recommend based on established knowledge obtained in laboratory settings. It seems relatively clear that the demands change with changes in the complexity of a decision as well as that adapting the decision mode to changes in situational complexity is very likely a good way to choose the most appropriate decision mode for a particular situation. In addition, the inverse U-shape that inexperienced decision-makers apply is a solid solution to many situations except the most complex decisions, for which we would encourage attempts to use a deliberate mode as far as possible.

This, in turn, implies that highly experienced decision-makers, who do not adjust their decision modes significantly to changes in subjective situational complexity, likely do not utilize their full potential. This is most obviously a problem for low-complexity decisions, for which routine-based intuition would be a highly efficient strategy, especially for experienced decision-makers with a strong set of expert schemas. Instead, we consistently see that highly experienced decision-makers strongly rely on deliberate decision-making for this type of decision. We would encourage a stronger use of a more intuitive decision mode for low-complexity decisions. When complexity increases, highly experienced decision-makers use comparatively more intuition than less experienced decision-makers (or at least not much less, the specific picture varies somewhat across the studies). This is appropriate, considering that highly experienced decision-makers have stronger expert schemas and can therefore leverage the benefits of a relatively intuitive mode better than less experienced decision-makers. Considering this, it would make sense to rely even more on intuition in this complexity range. At the high end of the complexity spectrum, highly experienced decision-makers are, just as less experienced decision-makers, well advised to use the intuitive mode to only a limited degree. This is oftentimes the case.

In summary, we believe that decision-making could be improved by encouraging highly experienced decision-makers to use more intuition for low-complexity decisions. At the same time, we would encourage less experienced decision-makers to use more deliberation for highly complex decision situations.

6.3 DEVELOPING A DECISION ENVIRONMENT THAT SUPPORTS EFFECTIVE DECISION-MAKING

Based on the discussion so far, it is time to think about what trainers and those responsible for organizational structures could do to improve decision-making. We think there are two approaches toward this goal: specific training in how to choose an appropriate decision mode and general organizational development strategies that help build a beneficial decision environment. We will begin the following discussion with suggestions for decision training.

6.3.1 RECOMMENDATIONS FOR DECISION-MAKING TRAINING

The first approach to include the insights developed in this book into the every-day practical work of professionals in complex task organizations is to improve decision-makers' sensitivity for different ways to approach decisions with the aim to help them select an appropriate one. Trainers would, in such a training, educate decision-makers such that they understand how they make decisions and then help them find the right decision mode for different situations while considering the advantages and disadvantages of intuitive and deliberate decision modes. Such an approach would imply implementing what we discussed in Sections 6.1 and 6.2 into a training program, either as an independent training approach or as part of an established training setup with the goal to help decision-makers develop self-awareness of the decision modes they use. Decision-making training would also help organizations and decision-makers overcome the preconception that deliberation, as the "rational" decision mode, is the "good" approach, whereas intuition is "irrational" or "bad." We have strong evidence, both from our own and from others' research, that this simplification is wrong: The modes are neither good nor bad; and which mode is more or less appropriate depends on the situation. Understanding these optimal conditions could move decision-making performance substantially forward.

As outlined above, there is no clear-cut recommendation that would hold consistently for any situation. However, there are recommendations that are quite generalizable based on the conclusions derived so far. Those could be broken down into relatively simple rules of thumb for choosing an appropriate decision mode. Such rules of thumb could be formulated as a series of simple questions that a decision-maker could employ to consciously inform his or her choice of decision modes. The following questions are examples of this approach:

- Is the decision set in a task context with which I am familiar? If yes, use intuition, otherwise deliberation.
- Have I often been in comparable situations before? If yes, use intuition, otherwise deliberation.
- Does the decision require more information than I currently have? If yes, use deliberation, otherwise intuition.
- Is the number of cues large or are cues difficult to perceive? If yes, use intuition, otherwise deliberation.

- Does the decision involve conflicting goals or unclear priorities? If yes, use deliberation, otherwise intuition.
- Am I overwhelmed with inputs? If yes, use intuition, otherwise deliberation.

These recommendations could be used to establish a decision mode tendency. That is, if a decision-maker arrives at two recommendations to use intuition and one to use deliberation in a given situation, he or she could conclude that a decision mode employing a substantial amount of intuition and a modest amount of deliberation would be a good choice.

As an example, envision a coordination task that a local incident coordinator might face, comparable to the task we saw in the first case study presented in Chapter 5. The decision-maker would be on board a distressed vessel, tasked with organizing a variety of actors on board the ship conducting medical tasks, evacuation, and technical assistance in parallel. At the same time, his or her job would also involve communicating with the captain of the vessel in distress and relaying information to the overall on-scene coordinator. This situation puts the decision-maker into a series of decision situations he or she must manage. Let us begin with the initial resource deployment decisions, assuming that the weather is good and the temporal organization on board is just unfolding. The first question from the list above is if the task context is familiar. As the on-scene coordinator will probably be trained in coordinating resources and be familiar with the general challenges he or she has to expect (such as that burning cargo might create poisonous fumes), the answer to this question is yes; and we have an initial tendency to use intuition. The second question is does the decision-maker have experience with that specific situation? If this is not the case, this would imply a tendency toward deliberation. If the decision-maker also lacks important information, he or she would have to obtain additional information from outside sources, for example, regarding the estimated time of arrival of additional resources on scene. This would again imply a tendency to use deliberation. As the weather is good and communication channels are functional, the relevant cues are not hard to perceive; this indicates a tendency toward deliberation. We assume that, at the beginning, he or she would also not be overwhelmed by the information input, leading to another recommendation to use deliberation, as well as that goals are not conflicting, leading to a recommendation to use intuition. Summing this up, we would see four recommendations to use deliberation and two for intuition. We would then recommend that the decision-maker use a relatively deliberate and structured approach, for example, writing down available resources and communication channels on a sheet of paper or a whiteboard (as available), checking with and acquiring information from the overall on-scene coordinator, and then developing an ideal plan for resource deployment.

Let us continue with this scenario a bit further. We assume that, during the assignment, the weather turns bad and the distressed vessel is now in heavy sea. In addition to ongoing operational tasks with medical aid, firefighting, and evacuation, the ship now also takes on water and begins to list. Due to heavy sea and a general state of anxiousness on board, radio communication becomes difficult; everyone on board attempts to get the attention of the local incident coordinator as well as obtain access to the available resources, which are already stretched. The decision-maker knows that he or she will not get outside help for some time. Under these conditions, the situation has changed. Cues are now, due to the noise, turmoil, and heavy sea state, difficult to perceive. The decision-maker feels that he or she is about to be overwhelmed by inputs. On the other hand, goals are now conflicting, as many are rivaling for scarce resources and there is no longer a need to acquire outside information. Based on the checklist of recommendations, this now leads to a situation in which we have four recommendations to use intuition and only two to use deliberation. At this point, it is now time to abandon the strongly structured approach and put trust in more holistic thinking, knowing that especially the lack of experience with this type of situation makes fully intuitive decisions somewhat risky. Thus, a good approach in this situation would be to take in as many cues as possible and decide quickly based on the information available but still grounded in the analytical steps taken previously, for example, by making continuous use of the artifacts employed previously, such as notebooks or whiteboards; retaining control over highly critical aspects, such as resource bottlenecks, and critical goals, such as the number of people on board versus the number of people saved; but otherwise speeding up decision-making. This more intuitive approach would free up cognitive resources for communication, reducing the bottleneck problem that the local incident coordinator is in danger of creating in this situation.

Oftentimes, the recommendations derived in this way will likely match the decision mode that a proficient decision-maker would choose anyway. However, as discussed in Section 6.2, there are a variety of instances in which this is not the case. For those cases, we believe that a decision mode–oriented training could potentially improve decision-makers' task performance.

The choice of decision mode is also a decision and one on which the decision-maker would typically not want to spend too much time and resources—that is, the choice of decision mode is a decision that should be made intuitively. This, in turn, implies that decision-makers should and must practice finding the right decision mode in various situations, just as they would practice any other skill they intend to move to the domain of routine-based intuition. A list of questions, such as the ones suggested earlier, would be a simple way to practice this skill. Developing intuition regarding when to use which decision mode should be relatively easy when relying on such simple guidelines. Yet, it still requires a certain amount of practice to memorize which conditions favor which decision mode.

For setting up training on decision-making, it might be helpful to start with filming the decision behavior of decision-makers. We used the approach of video-based shadowing studies to collect data for this book for scientific purposes. However, we found that such video footage also has strong instructive value, as it encourages decision-makers to reflect on their decision behavior and thus develop the self-awareness necessary to improve decision performance. Later on, and based on this introspection, decision-makers would need to be taught thoroughly how the specific decision modes work.

In summary, decision mode–oriented training could look as follows. Trainees would be videotaped in training environments that reflect challenging decision situations of varying complexity. The instructor would then watch the video footage together with the trainees and discuss the decision mode chosen as well as how the trainee experienced those decisions; the trainees should develop a certain understanding of the different approaches they can make or could have made as well as how these "feel" cognitively: Do they imply a heavy load on cognitive capacities or not? Are they fast or slow? Analytical or not? And so on. In the next stage, the trainer would familiarize the trainees with the two archetypical decision modes as well as their specific advantages and disadvantages. Afterward, decision-makers should discuss different decision situations and come up with reasons to use one or the other archetype for each. The end of this process could be a list of questions comparable to the one suggested earlier, which trainees could use, refine, and practice with. In the following, this list, if it makes sense to the trainees and proves to be useful, should be practiced continuously until selecting the appropriate decision mode becomes part of work routines. The initial part of such training could be completed somewhere between three hours and one day, dependent on whether the video material is already available prior to the training session. The training could, for example, be integrated into the debriefing of a larger exercise or conducted independently in a simulator environment. A series of follow-up refreshers would then help decision-makers bring the list of questions to the routine level.

To date, there are very few ideas on training for how to choose an appropriate decision mode in complex task environments. Instead, most approaches focus on "debiasing," which is, in the end, a way to move decision-making toward more deliberation. As we learned, this is not always the best choice. We believe that a good training approach should build on and leverage the benefits of both intuitive and deliberate decision modes. The training approach outlined earlier, in contrast, implies practicing the choice of decision modes without preferring one over the other.

We discussed training the choice of decision modes earlier. Yet is there a way to practice the more effective use of a given decision mode? The answer to that question is rather trivial: There certainly is, but this is already being done in standard trainings. Making better decisions with an intuitive mode means improving expert schemas, which develop through repeated exposure to certain decision situations. On the other hand, making better deliberate decisions primarily involves the development of broad, analytical, and partially abstract knowledge in addition to developing work routines that help group, weigh, and select cues and decision rules. This is the type of development "traditional" trainings, whether in a real-life setting or a simulator or practice environment, are meant to achieve anyway. Thus, we believe that decision training on the individual level has much to do with skill development in the form it is being applied today. This underlines the critical importance of established training practices for successful performance in complex task environments (see also Steigenberger, 2016).

On the organizational level, however, we see much room for improvement. We believe that it is possible and likely helpful to set work spaces and organizational environments up in such a way that they facilitate the development of expert schemas and analytical capabilities and allow individuals to choose the appropriate decision mode more effectively than it is being done today. We discuss this in the following.

6.3.2 Recommendations for Organizing Complex Task Environments

Decision-makers are embedded in organizational and situational environments that determine how they act as well as how effectively they conduct their work.

It is therefore important to consider the organizational environment when considering how effective decision-making can be facilitated. Three dimensions need to be considered: Setting up good environments for deliberate decision-making, setting up good environments for intuitive decision-making, and finally, setting up environments such that they allow for a flexible application of deliberate or intuitive decision-making.

Let us begin with the first point, preparing organizations for effective deliberate decision-making. In the Western world, we already have a long tradition of facilitating deliberate decision-making. Accordingly, many organizations in complex task environments are already set up such that they effectively support deliberation. One aspect of this is a relatively broad and general education that focuses on developing abilities and competences in abstract thinking and purposeful knowledge selection and rule application. In combination with education that provides generalizable insights in the form of lectures or seminars, these approaches build the required capabilities for deliberation. Education is limited in this respect to some degree as the abilities to use deliberation are linked to concepts of general intelligence (Evans, 2006). Otherwise, the current concept of educating professionals in complex task environments is suited to improve deliberate decision-making capabilities.

For setting up work environments that support deliberation, there is an ongoing debate on decision support systems, which have the main intention of reducing the cognitive load of the deliberate mode and thus increasing the applicability and accuracy of deliberate decision-making. For example, on the behavioral side, Jenkins, Stanton, Salmon, Walker, and Rafferty (2010) suggest the "decision ladder" as a set of different stages that a decision-maker should go through when approaching a decision situation. In the field of technical solutions, Kapucu, Augustin, and Garayev (2009) suggest network analysis software to aid complex coordination decisions and related learning processes, whereas Nordström et al. (2016) provided a "vessel TRIAGE" tool to aid status assessments. Finally, Crichton and Flin (2001) suggest simulation games as a method to develop analytical skills for rare events. It is also important to note that even very simple physical tools help tremendously to overcome the cognitive load problem; writing important information down on a whiteboard, for example, can make all the difference when cues are coming in fast and decision-makers are in danger of forgetting or misinterpreting important inputs. It is beyond the scope of our book to delve deeper into the wide field of decision support systems for deliberate decision-making. Yet many solutions are already being discussed that help decision-makers to apply the deliberate mode effectively. Especially the simple solutions proved to be very helpful in that respect.

The situation looks somewhat less positive for the intuitive decision mode. First, intuition is not related to general intelligence; and effective intuition is very situational. This means that the only way to improve the capability to use an intuitive mode effectively is to build expert schemas by being exposed to the specific situations to which intuition relates; lectures and seminars will not do much good in this respect. This is likely not a problem for routine-based intuition—being exposed to everyday assignments in combination with regular trainings of somewhat less frequent occurrences—will provide the necessary exposure here. However, at least in the field we studied, moderate to highly complex decisions occur very infrequently.

It is therefore extremely difficult to build expert schemas for this type of task. Intuition, on the other hand, would be useful as the downsides of deliberation are most critical for highly complex decisions.

What could possible solutions be? The most obvious is to put a heavy emphasis on training environments that allow practicing high-complexity decisions in a realistic environment. Studies 3 and 4 in Chapter 4 and case study 1 in Chapter 5 relate to training environments that satisfy this condition (see Chapter 3 for background). Yet such trainings are very expensive, both in terms of financial cost and commitment from senior trainers. These costs make it unrealistic to expect that such trainings could be provided regularly for all employees in complex task environments.

A more economical solution would be an advanced simulator environment. Simulators have the immanent problem that they do not put decision-makers in the same situations as real-life exercises would—there is no real danger; the environment is artificial; and oftentimes compromises in terms of realism of the training hardware must be made so that, for example, not all tools available in real-life are realistically simulated. Training will therefore be less effective in terms of building expert schemas for real-life situations compared to real-life exercises. Still, effective simulator environments might still be a realistic alternative for many organizations, in particular if newer options in simulator technology are being used, such as having spatially distributed simulators linked together into a joint complex scenario. New developments, such as virtual reality technology, might also contribute to improving simulator learning in terms of developing expert schemas. Still, developing intuition without continuous and repeated exposure to complex decision situations in real-life environments will always be difficult.

When designing training environments that should help decision-makers develop expert schemas, some other considerations need to be kept in mind, as well. In particular, it has been noted that trainings and practical exposure can also create false learning in what has been called "wicked" learning environments (Hogarth, 2005). For effective learning, the environment must provide stable and visible feedback to the decision-maker. That is, a particular decision should consistently lead to a particular outcome, and the decision-maker needs to receive the related feedback. If variance in outcomes is high—that is, if the chance that, due to conditions outside of the control of the decision-maker, a suboptimal decision leads to a positive outcome and vice versa—false learning is likely to occur. In this case, decision-makers unwarrantedly disregard high-quality decisions because the outcome is negative or stick to low-quality decisions when the outcome is positive. This form of outcome bias might then lead decision-makers to develop expert schemas that are actually wrong and which will guide them to poor intuitive decisions. If the environment is such that the decision-maker does not receive feedback at all, he or she cannot develop expert schemas for this environment. Both problems are also relevant for the training of deliberate decision processes, yet as the effectiveness of intuition rests so centrally on accurate and rich expert schemas, their impact on intuitive decision modes is most severe.

Such wicked learning environments occur often, for example, in business firms, where it is extremely difficult to tell whether a positive or negative market outcome is the result of a particular management decision or whether other conditions affected or caused this outcome (Zollo, 2009). However, many safety-oriented organizations operate in stable and relatively predictable task contexts, in which causal ambiguity is less of a problem. Still, this issue needs to be kept in mind; in "wicked" environments, developing strong and reliable expert schemas is much more difficult and requires much more exposure to the situation to which the expert schemas relate (see also Kahneman & Klein, 2009). Expert schemas are critical for effective decision-making and, thus, highly important for intuitive decision-making. Hence, organizations are well advised to develop expert schemas in their employees.

Finally, decision-makers should also be able to choose the decision mode most appropriate for a certain situation. Organizational embedment can have a substantial bearing on whether or not this is the case. In particular, organizational practices might implicitly or explicitly encourage a particular decision mode, making it less likely that decision-makers choose the less favored mode even in situations in which the less favored mode would be better. For example, we found that in maritime SAR, the field we studied for this book, there is an emphasis on deliberate decision-making, whereas intuitive modes are discouraged to some degree. This emphasis is relatively subtle and manifested, for example, in the use of words: Deliberate decision-making is referred to as "rational," which carries a positive connotation, whereas intuition is referred to as "feeling," which does not carry a negative connotation per se but is clearly less positive than the "rational" way to approach a decision. The structure of debriefings further complicates intuitive decision-making, as we discussed in this book on various occasions. Intuitive decision-makers typically cannot recall the rules they used when making a decision, so they find it much more difficult to defend their actions in debriefing situations, in which decision-makers might be asked to explain why a specific decision has been made. This problem is somewhat mitigated by an often relatively friendly and trustful environment in these debriefings as well as by the abilities of decision-makers to retrospectively rationalize their decisions even if made intuitively (see Catino & Patriotta, 2013).

So how could organizations be made fit for flexible choices of decision modes and the use of intuition? We believe that the answers are relatively simple or maybe even disappointing. Intuition rests squarely in the subconscious of the decision-maker. Thus, organizations cannot do much to foster this type of decision-making aside from helping decision-makers develop expert schemas and increasing the tolerance for intuitive decision-making in their general perspective of and communication about intuition. It is therefore not so much about encouraging intuition; considering the downsides of immature intuition, organizations should consider carefully if they should move in that direction. It is instead about not discouraging intuition, provided that intuitive decision-makers have the necessary expert schemas that turn intuition into a powerful decision tool for many situations. Methods to do that are to increase the psychological safety through trustful relationships between employees and, especially, leaders and followers and to instill an error and safety culture that accepts that deliberate as well as intuitive decision modes will produce errors (Catino & Patriotta, 2013). Tools are nonpunitive approaches to errors, safe working climates, and trust building. Achieving this is a leadership challenge for organizations that act in complex task environments.

6.4   OPEN QUESTIONS, LIMITATIONS, AND FINAL WORDS

The approach taken in this book has several limitations, and it is appropriate to discuss them at this point in the book. There are, in particular, two empirical limitations that should be noted. First, as we cannot look into the minds of the decision-makers during real-life assignments, we must rely on self-reports of decision modes (for the survey studies) or our own or expert interpretations of observed events (for the video-based studies). This approach introduces a source of error, as decision-makers do not respond immediately after the decision to the survey questions but, rather, sometime later when the assignment is already over. It is conceivable that decision-makers rationalize their decision modes up to that point or that they simply forget exactly how they made a decision. This retrospection bias is a general problem in this type of research, which we addressed in three ways: First, we attempted to minimize the time between decision and response in our study design. Second, we used scales with a variety of items, which are more robust to retrospection bias than single item measures would have been. For the video-based studies, interpreting the observations with deep field knowledge as well as cross-checking interpretations with field experts helped us ward off the danger of false interpretations. Third, we combined different studies and measures to increase the robustness of the results. Reproduction bias leads to measurement errors, which introduce random variance to the results. This means that measurement error would make it less likely that results are reproduced over different studies. As our core results do reproduce despite measurement error, we are quite confident that they stand on strong empirical ground.

The second empirical problem is the interpretation of results in the face of causal ambiguity. Based on our measures in the survey studies, we capture the co-occurrence of observations, but it is beyond the abilities of our statistical approach to establish causality. We see, for example, that the relationship between situational complexity and the use of deliberation is inverse U-shaped, but we cannot determine with certainty whether the change in situational complexity actually caused this shape. Considering that we rule out several alternative explanations by including the respective alternative measures as variables in the statistical models and also considering that we have solid theoretical reasons to assume that causality exists, we are confident that the claims we make do hold. However, future research is needed to establish these claims. Again, replication helps us strengthen our confidence in the causal relationships we interpret into our observations. As our core results replicate over studies, we are reasonably sure that the relationships we found are empirically and conceptually sound.

Having solid solutions for the empirical limitations, we are quite certain that our results provide a strong and stable foundation for advancing our understanding of decision-making in complex task environments. We provide rich empirical insights and hope that these, in combination with the recommendations for training and organizational development we provide, are insightful for researchers as well as field practitioners. Yet our results also opened up several new or follow-up questions that provide promising directions for future research.

In particular, we would encourage researchers in future studies to continue our examination of how decision-maker characteristics affect the choice of decision mode. This is a topic that features prominently in laboratory research but has been largely neglected by field research in the tradition of the "naturalistic decision-making" stream. In our study, we examined biographical data, mindfulness, decision mode preference, and stress resistance. It is conceivable that other personality characteristics that have been shown to affect work performance in different settings also matter for the choice of decision modes. For example, commitment and identification (Knippenberg, 2000; Neininger, Lehmann-Willenbrock, Kauffeld, & Henschel, 2010) as well as the "Big Five" personality traits (openness, emotional stability, extroversion, agreeableness, and conscientiousness, see Judge & Zapata, 2015) could easily also have an impact on how decision-makers approach complex judgmental decisions. Another open venue in this line of thinking is the role of emotions. We know that particularly type 1 processing is strongly affected by emotions (Dane & Pratt, 2007). We only tangentially addressed this topic in this book. Yet, future research might look more closely into how emotions in general and specific emotions shape the choice and effectiveness of decision modes, linking studies of the type we presented here with previous work done in various fields of psychology.

Then, it is important to note that we studied decision-making in a specific setting: high-reliability organizations (see LaPorte, 1996; Roberts, 1990; Vogus & Welbourne, 2003, for a description). High-reliability organizations are governed by certain characteristics that are not found in many other organizations, such as a focus on error avoidance and safety as an organizational goal or relatively stable technical systems. Although we believe that the approach to judgmental decisions under complexity should be relatively independent of these organizational characteristics, it would be helpful to have empirical insights regarding the degree to which our findings replicate in fields, such as the management of traditional business firms. We discussed some lines of thought in this direction throughout the book, yet it was beyond our focus to develop these questions in detail.

Finally, and perhaps most importantly, in this book we examined situations in which decisions are made mostly by individuals, embedded in a relatively strong hierarchy that clearly prescribes roles and responsibilities. In many organizations, decisions are made in other forms and roles are oftentimes negotiated or fluent—that is, they develop during social interactions (Bechky, 2006). Team decision-making or fluent and unstable roles are important in, for example, top management teams in business firms or ad hoc organizations in the high reliability field (see, e.g., Bromiley & Rau, 2016; Wolbers & Boersma, 2013). In these situations, team members can deliberate or intuit on a decision problem, but the decision is ultimately made after discussing the different ideas for a decision within the team. We currently have very little understanding of how team processes, team composition, and team task characteristics affect the choice of decision modes of team members as well as how this, in turn, affects the decisions that teams ultimately make. The second case study presented in Chapter 5 provides some indications that team processes might be quite important for the establishment of team decision routines as well as that such routines, in order to develop, require continuous repetition (see also Feldman & Pentland, 2003). It is, for example, conceivable that intuition is rare in teams where team members do not trust each other, as such teams would require team members to defend their decisions in a potentially unfriendly environment. This is much more difficult for decisions made with an intuitive mode than for decisions made with a deliberate mode (see Section 2.3.2). This would force teams characterized by low levels of mutual trust to use comparatively more deliberation, which, in turn, could lead to unwarranted cognitive opportunity costs (Kurzban, Duckworth, Kable, & Myers, 2013) and thus hamper team performance. Another twist in that direction would be to look at how intuitive and deliberate processes on the individual level affect the decision-making process of a team. That is, whether opinions developed in an intuitive way influence the decisions of teams more or less toward the opinion of the intuitive decision-maker as compared to opinions developed in a deliberate way. In addition, one could examine which characteristics affect this process. Decision modes in teams are a topic that would help us better understand how decisions emerge from and in teams.

Overall, we believe that our book answered various important questions and helped us understand how decision-makers decide in complex task environments. However, developing knowledge on such a complex topic is a cumulative process. We built on a strong conceptual and empirical foundation and body of knowledge and contributed our humble insights to this ongoing discourse. Our hope is that this book can be a modest building block for the continuous development of our shared attempt to understand how humans make decisions in real-life environments as well as help practitioners realize how decision-making can be improved in order to steadily increase the safety, efficiency, and humaneness of important organizations in our societies.

REFERENCES

Alter, A. L., Oppenheimer, D. M., Epley, N., & Eyre, R. N. 2007. Overcoming intuition: Metacognitive difficulty activates analytic reasoning. *Journal of Experimental Psychology. General*, 136(4): 569–576.

Baylor, A. L. 2001. A U-shaped model for the development of intuition by level of expertise. *New Ideas in Psychology*, 19(3): 237–244.

Bechky, B. A. 2006. Gaffers, gofers, and grips: Role-based coordination in temporary organizations. *Organization Science*, 17(1): 3–21.

Betsch, C. 2008. Chronic preferences for intuition and deliberation in decision making: Lessons learned about intuition from an individual differences approach. In H. Plessner, C. Betsch, & T. Betsch (Eds.), *Intuition in judgement and decision making* (pp. 231–248). New York, London: Lawrence Erlbaum Associates Inc.

Boin, A., & Schulman, P. 2008. Assessing NASA's safety culture: The limits and possibilities of high-reliability theory. *Public Administration Review*, 68(6): 1050–1062.

Bromiley, P., & Rau, D. 2016. Social, behavioral, and cognitive influences on upper echelons during strategy process: A literature review. *Journal of Management*, 42(1): 174–202.

Catino, M., & Patriotta, G. 2013. Learning from errors: Cognition, emotions and safety culture in the Italian Air Force. *Organization Studies*, 34(4): 437–467.

Crichton, M., & Flin, R. 2001. Training for emergency management: Tactical decision games. *Journal of Hazardous Materials*, 88(2–3): 255–266.

Dane, E. 2010. Reconsidering the trade-off between expertise and flexibility: A cognitive entrenchment perspective. *Academy of Management Review*, 35(4): 579–603.

Dane, E., & Pratt, M. G. 2007. Exploring intuition and its role in managerial decision making. *Academy of Management Review*, 32(1): 33–54.

Evans, J. S. B. T. 2006. The heuristic-analytic theory of reasoning: Extensions and evaluation. *Psychological Bulletin & Review*, 13(3): 378–395.

Feldman, M. S., & Pentland, B. T. 2003. Reconceptualizing organizational routines as a source of flexibility and change. *Administrative Science Quarterly*, 48(1): 94–118.

Hensman, A., & Sadler-Smith, E. 2011. Intuitive decision making in banking and finance. *European Management Journal*, 29(1): 51–66.

Hogarth, R. M. 2005. Deciding analytically or trusting your intuition? The advantages and disadvantages of analytic and intuitive thought. In T. Betsch & S. Haberstroh (Eds.), *The routines of decision making* (pp. 67–83). Mahwah, NJ: Lawrence Erlbaum Associates Inc.

Hollnagel, E. 2006. *Resilience engineering: Concepts and precepts*. Aldershot: Ashgate.

Jenkins, D. P., Stanton, N. A., Salmon, P. M., Walker, G. H., & Rafferty, L. 2010. Using the decision ladder to add a formative element to naturalistic decision-making research. *International Journal of Human-Computer Interaction*, 26(2–3): 132–146.

Judge, T. A., & Zapata, C. P. 2015. The person–situation debate revisited: Effect of situation strength and trait activation on the validity of the big five personality traits in predicting job performance. *Academy of Management Journal*, 58(4): 1149–1179.

Kahneman, D., & Frederick, S. 2002. Representativeness revisited: Attribute substitution in intuitive judgment. In T. Gilovich, D. Griffin, & D. Kahneman (Eds.), *Heuristics and biases: The psychology of intuitive judgment* (pp. 49–81). Cambridge University Press.

Kahneman, D., & Klein, G. 2009. Conditions for intuitive expertise: A failure to disagree. *American Psychologist*, 64(6): 515–526.

Kapucu, N., Augustin, M.-E., & Garayev, V. 2009. Interstate partnerships in emergency management: Emergency management assistance compact in response to catastrophic disasters. *Public Administration Review*, 70(2): 297–313.

Klein, G., Calderwood, R., & Clinton-Cirocco, A. 2010. Rapid decision making on the fire ground: The original study plus a postscript. *Journal of Cognitive Engineering and Decision Making*, 4(3): 186–209.

Klein, G. A. 1995. A recognition-primed decision (RPD) model of rapid decision making. In G. A. Klein, J. Orasanu, R. Calderwood, & C. E. Zsambok (Eds.), *Decision making in action. Models and methods* (2nd ed.) (pp. 138–148). Norwood: Ablex Publishing.

Klein, G. A., Orasanu, J., Calderwood, R., & Zsambok, C. E. (Eds.) 1995. *Decision making in action: Models and methods* (2nd ed.). Norwood: Ablex Publishing.

Knippenberg, D. van 2000. Work motivation and performance: A social identity perspective. *Applied Psychology*, 49(3): 357–371.

Kurzban, R., Duckworth, A., Kable, J. W., & Myers, J. 2013. An opportunity cost model of subjective effort and task performance. *The Behavioral and Brain Sciences*, 36(6): 661–679.

LaPorte, T. R. 1996. High reliability organizations: Unlikely, demanding and at risk. *Journal of Contingencies and Crisis Management*, 4(2): 60–71.

Lipshitz, R., Klein, G., Orasanu, J., & Salas, E. 2001a. A welcome dialogue—And the need to continue. *Journal of Behavioral Decision Making*, 14(5): 385–389.

Lipshitz, R., Klein, G. A., Orasanu, J., & Salas, E. 2001b. Taking stock of naturalistic decision making. *Journal of Behavioral Decision Making*, 14(5): 331–352.

Neininger, A., Lehmann-Willenbrock, N., Kauffeld, S., & Henschel, A. 2010. Effects of team and organizational commitment—A longitudinal study. *Journal of Vocational Behavior*, 76(3): 567–579.

Nemeth, C., & Klein, G. 2011. The naturalistic decision making perspective. *Wiley Encyclopedia of Operations Research and Management Science*.

Nordström, J., Goerlandt, F., Sarsama, J., Leppänen, P., Nissilä, M., Ruponen, P., Lübcke, T., & Sonninen, S. 2016. Vessel TRIAGE: A method for assessing and communicating the safety status of vessels in maritime distress situations. *Safety Science*, 85: 117–129.

Roberts, K. H. 1990. Some characteristics of one type of high reliability organization. *Organizational Science*, 1(2): 160–176.

Steigenberger, N. 2016. Organizing for the big one: A review of case studies and a research agenda for multi-agency disaster response. *Journal of Contingencies and Crisis Management*, 24(2): 60–72.

Thompson, V. A., Turner, J. A. P., Pennycook, G., Ball, L. J., Brack, H., Ophir, Y., & Ackerman, R. 2013. The role of answer fluency and perceptual fluency as metacognitive cues for initiating analytic thinking. *Cognition*, 128(2): 237–251.

Tversky, A., & Kahneman, D. 1979. Prospect theory: An analysis of decision under risk. *Econometrica*, 47(2): 263–292.

Tversky, A., & Kahneman, D. 2000. Judgment under uncertainty: Heuristics and biases. In T. Connolly, H. Arkes, & K. R. Hammond (Eds.), *Judgment and decision making: An interdisciplinary reader* (pp. 35–52). Cambridge: Cambridge University Press.

Vogus, T. J., & Welbourne, T. M. 2003. Structuring for high reliability: HR practices and mindful processes in reliability-seeking organizations. *Journal of Organizational Behavior*, 24: 877–903.

Wolbers, J., & Boersma, K. 2013. The common operational picture as collective sensemaking. *Journal of Contingencies and Crisis Management*, 21(4): 186–199.

Zollo, M. 2009. Superstitious learning with rare strategic decisions: Theory and evidence from corporate acquisitions. *Organization Science*, 20(5): 894–908.

Index

Page numbers followed by f and t indicate figures and tables, respectively.